Other Titles in the Aldus **Modern Knowledge** Series:

Man Probes the Universe *Colin A. Ronan*
Treasures of Yesterday *Henry Garnett*
Explorers of the World *William R. Clark*
World Beneath the Oceans *T. F. Gaskell*
Animal Behaviour *J. D. Carthy*
Through the Microscope *Margaret Anderson*
Men without Machines *Cottie Burland*
Man and Insects *L. Hugh Newman*
Earth's Crust *Keith Clayton*
Energy into Power *E. G. Sterland*
Man Alive *G. L. McCulloch*
Zoos of the World *James Fisher*
The Air around Us *T. J. Chandler*
Man, Nature and History *W. M. S. Russell*
Fresh Water *Delwyn Davies*

Man and Plants

H. L. Edlin

Man and Plants

Aldus Books London

Editor Kit Coppard
Designer Arthur Lockwood
Assistant Edward Poulton
Research Rosemary Barnicoat
Patricia Quick
Judy Aspinall

ISBN: 0490 00078 9

First published in 1967 by
Aldus Books Limited
Aldus House, Fitzroy Square, London W1
Distributed in the United Kingdom
and the Commonwealth by
W. H. Allen & Company
43 Essex Street, London WC2

Printed in Italy by Arnoldo Mondadori Verona

Contents

1 Plants—the Primary Food Makers

The origin of life is one of the most fascinating but difficult problems in science. Although it has not yet been solved, we do have an idea as to when and under what conditions the first life-forms appeared on this planet. It is probable that these life-forms—bacteria and other single-celled organisms—came into being a little less than 3000 million years ago. Such life functioned without oxygen because this gas, which now forms about one fifth of our atmosphere, was at that time almost entirely absent. It seems certain, too, that the organisms were confined to the bottom of ponds or shallow lakes, where the water was deep enough to protect them from certain lethal types of solar radiation and yet was not too deep to deprive them of other rays that were needed for growth.

During the next 2000 million years or so these primitive organisms no doubt diversified, but none of them made any significant climb up the evolutionary ladder. By about 420 million years ago, however, the situation had changed. Not only had the concentration of atmospheric oxygen increased, but the atmosphere had developed a protective layer that prevented the deadlier kinds of solar radiation from reaching the earth's surface. Quite suddenly, then, conditions were ripe for the appearance of new, higher forms of life. Some of these were the ancestors of modern plants.

It is impossible to exaggerate the importance of plants to all the later biological events in earth's history. As we shall see, plants not only increased the supply of oxygen to the point where air-breathing animals could evolve; they also provided the organic matter on which animals could feed. Today, of the 300,000-odd species of plants that have been identified and named, only a small proportion are directly useful to man and to the animals he has domesticated. Yet, without plants, neither man nor any other animal could survive. Animals depend for nourishment on "ready-made," *organic* food; that is, on the substances out of which living matter is made. But where does "living matter" come from? The answer is: from

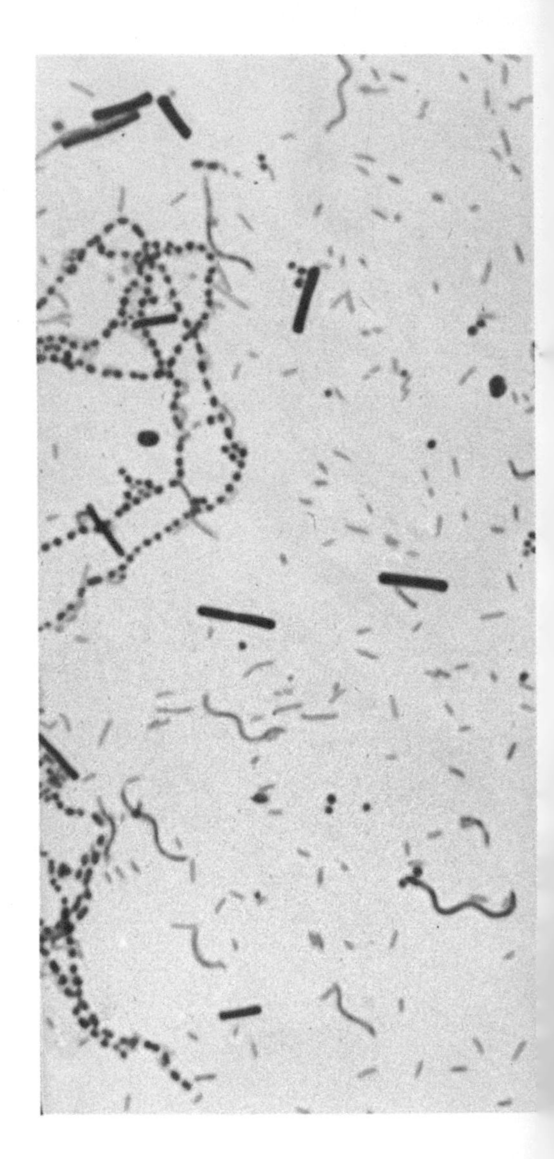

Bacteria (above) are among the smallest and most primitive forms of plant life, few species being able to synthesize food as do other plants. Photo shows three forms: spherical cocci, rod-like bacilli, and spirilla (× 640).

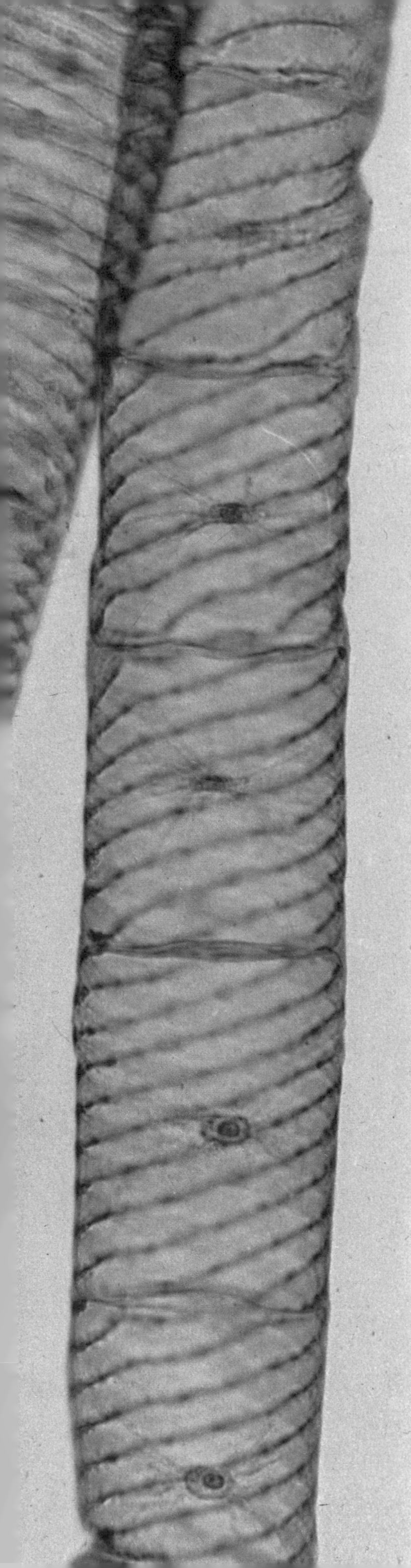

plants. Apart from bacteria and other primitive organisms, only plants are capable of manufacturing food from *inorganic* (non-living) matter. Thus, plants are the primary producers in nature—the essential starting point in the seemingly endless food chain that nourishes the myriad animal species on our planet.

Although we consider a wide range of plants in this book, much of our concern is with man's use of plants for food. But what exactly *is* food? We sometimes compare the role of food with that of a fuel that powers a machine. But we should not take this analogy too far. Fuels such as coal or oil are chemically very different from the materials of which the machine is constructed; food, on the other hand, consists of the same kinds of substances as those of our bodies. Food, in fact, has two jobs to do. First, it provides raw materials from which the body makes new tissues; second, like fuel, it releases energy that enables the body to function normally. There are many sources of energy on earth, but almost all of them are ultimately derived from the radiant energy that reaches us in the form of light from the sun. The physicist recognizes two main categories of energy: *potential energy*, which can be stored

The earliest oxygen-producing plants to evolve on earth may have resembled blue-green algae, which grow mainly in fresh water. The species above (×125) is *Aphanizomenon*. Left: another alga, *Spirogyra* (× 52), is commonly found on the surface of ponds. The spiral "ribbon" contains chloroplasts.

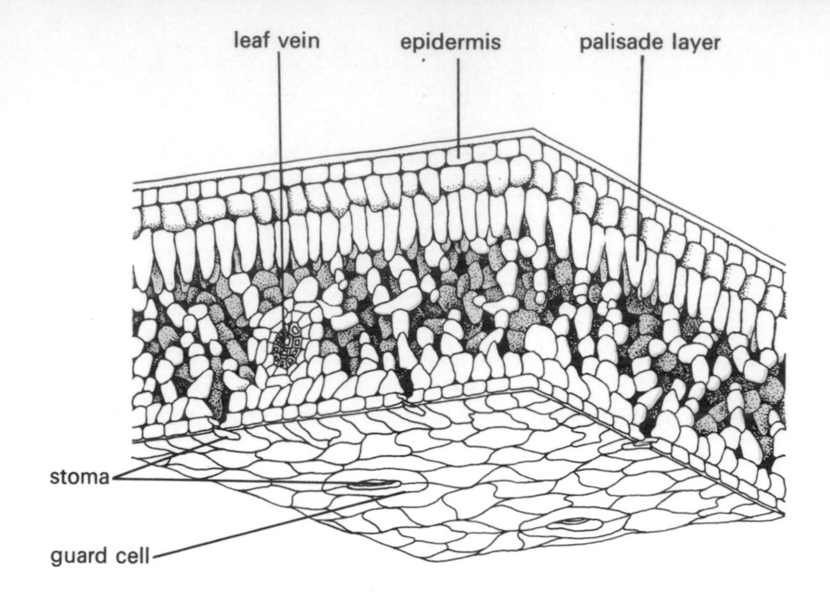

Plants synthesize food mainly in their leaves. The cross-section (left) of a typical leaf shows the stomata in the epidermis (leaf surface) through which carbon dioxide and water vapour pass into the inner leaf. Guard cells that surround the stomata control the volume of inflow. Photosynthesis, by which the raw materials are converted into glucose, occurs mainly in the palisade layer, the cells of which contain vast quantities of chloroplasts.

until it is required for use; and *kinetic energy*, the energy of movement, which must be used at once. Sunlight is a form of kinetic energy. A plant uses it to convert the raw materials of its food-making activities into potential energy in the form of chemicals that enable it to grow and reproduce itself.

The process by which plants produce food is called *photosynthesis* (from Greek words meaning "light" and "putting together"). There are probably at least 12 stages in the food-producing process: the initial stages, we know, depend upon sunlight; the later stages do not. The basic raw materials are carbon dioxide (CO_2), a gas that forms three parts of every 10,000 parts (by volume) of the atmosphere; and water (H_2O), which plants obtain from the soil.

Most plants, as we shall see, are able to take up water through their roots. But how is a plant able to absorb—to breathe in, as it were—the carbon dioxide in the air? The answer is that it does so through its leaves, where the photosynthetic process mainly occurs. The surface of a leaf contains countless millions of tiny openings, called *stomata* (singular, *stoma*—the Greek word for "mouth"), through which the gas is able to pass. The carbon dioxide, however, is destined for a particular part of each of the millions of cells of which the leaf is made. If we examine a plant cell under the microscope we find that its wall is lined on the inside with a thin layer of material. This is called the *cytoplasm*, and it contains the *nucleus*, which controls the physiological functions of the cell, including nourishment, growth, and cell division. Some of these functions are carried out within the cytoplasm by a variety of bodies called *plastids*. In the green tissues of plants (mainly in the leaves and other parts exposed to sunlight), some of the most important

The diagram (below) of a plant cell in cross-section shows the nucleus, which controls the activities of the cell, including its division. The plastids, some of which contain chlorophyll, are in the cytoplasm lining the cell wall. The fluid-filled vacuoles are concerned with storage and with waste removal.

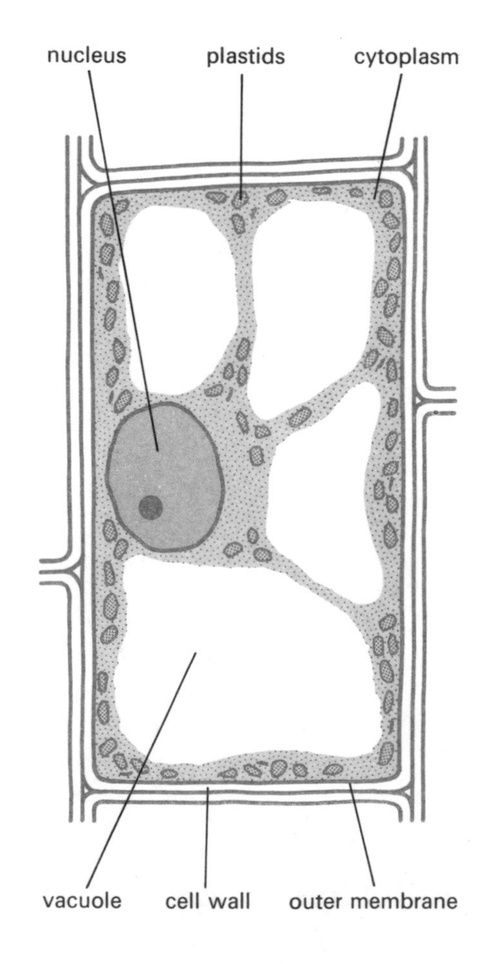

plastids are the *chloroplasts*—small, usually oval bodies within which photosynthesis takes place. Each chloroplast contains a green pigment called *chlorophyll* (from Greek words meaning "the green of the leaf") that plays a central role in photosynthesis.

As a matter of fact, several pigments are involved in photosynthesis. The four most important pigments are chlorophylls *a* and *b*, both of which are green in colour, and *xanthophyll* and *carotene*, which are orange-yellow. They are different in colour because they absorb light of different wave-lengths. Since photosynthesis is powered by the energy of sunlight, the capacity to absorb light of various wave-lengths is useful to a plant. The photochemical reactions are, in fact, initiated by chlorophyll *a* after it has collected most of the light energy absorbed by the other pigments. Chlorophyll is a compound of carbon, hydrogen, oxygen, nitrogen, and magnesium. Interestingly, its formula resembles that of haemoglobin, the substance that enables animals to release energy by carrying oxygen to the tissues.

Both carbon dioxide and water, then, must be present in the cytoplasm before photosynthesis can occur. The cytoplasm, in fact, consists of a high proportion of water that has seeped in through the cell wall. As the molecules of carbon dioxide pass through the stomata, they dissolve in the water at the outer surface of the cell wall and enter the cytoplasm by a process known as diffusion (see page 19).

Although the multi-stage process of photosynthesis is immensely complicated, it is solely concerned with converting carbon dioxide and water into *glucose*, the simplest form of sugar, which is an abundant store of potential chemical energy. The chemical formula of glucose is $C_6H_{12}O_6$. Thus, bearing in mind the formulas of carbon dioxide (CO_2), and water (H_2O), it is clear that the essential function of chlorophyll is to produce a series of reactions by which hydrogen atoms are removed from water, so breaking it down and releasing its other component, oxygen. Later, these hydrogen atoms are made to combine with the carbon dioxide to form glucose. As it happens, two molecules of water are needed for each molecule of carbon dioxide, so the reaction is as follows:

$$6CO_2 + 12H_2O + \text{light} \rightarrow C_6H_{12}O_6 + 6O_2 + 6H_2O$$

The oxygen ($6O_2$) that is one of the products of photosynthesis is in fact the principal source of atmospheric oxygen on our planet. It is essential to all the natural and "man-made" processes of combustion or oxidation that release energy.

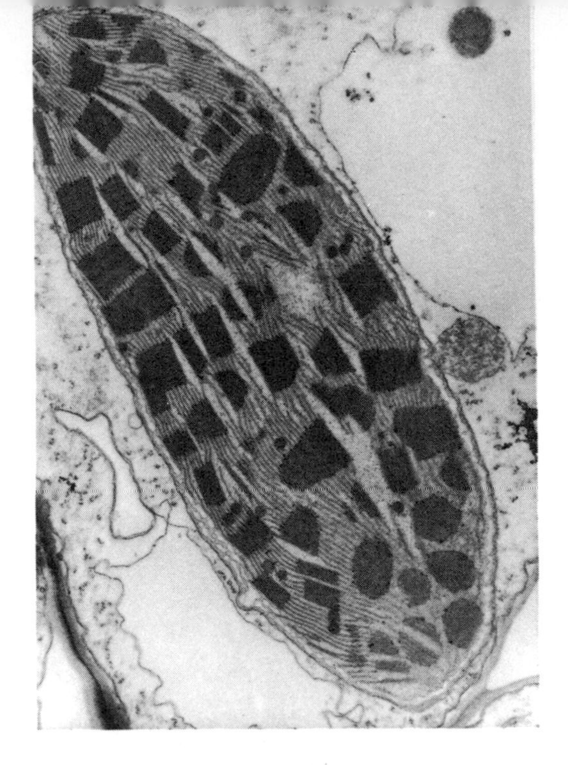

Electron micrograph (×10,000) of a maize-leaf chloroplast. Photosynthesis is believed to occur within the dark, angular units, or lamellae (seen here in cross-section). Each lamella contains several hundred chlorophyll molecules that trigger photosynthesis when they are energized by absorption of sunlight.

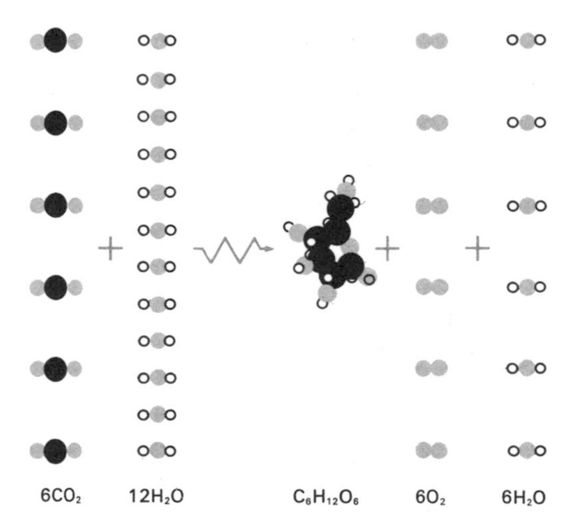

The multi-stage photosynthetic process is summarized above; carbon atoms are black, hydrogen atoms white, oxygen atoms green. The oxygen atoms in the glucose molecule derive from the carbon dioxide; those in the water molecules to the left of the glucose are released into the atmosphere in photosynthesis.

Food Types

Photosynthesis, then, is the basic process in a chain of events that provides plants, and hence animals, with all the food substances that they need for life and growth. The average European or North American has at his disposal an enormous variety of foodstuffs. But all of them contain one or more of three basic types of food: carbohydrates, fats, and proteins. Each of these basic types of food is synthesized by the plant from its store of the primary foodstuff, glucose.

Carbohydrates, which are the main source of energy in plants and animals, consist entirely of carbon, hydrogen, and oxygen. In the chemical formula of a carbohydrate, the hydrogen and oxygen atoms occur in the same ratio as they do in water; a carbohydrate has twice as many hydrogen atoms as oxygen atoms. There are three main types of carbohydrates: sugar, starch, and cellulose. Of the sugars, we are already familiar with glucose, the product of photosynthesis. Another is fructose, which has the same formula as glucose ($C_6H_{12}O_6$) and is abundant in fruits. A third is sucrose ($C_{12}H_{22}O_{11}$), which is formed by the combination of glucose and fructose and is the sugar we commonly use to sweeten our food.

Starches have much larger molecules than the sugars from which they are synthesized. They also differ from sugars in that they consist of grains of material, which vary in size according to the plant species in which they occur. Most plants convert their sugars into starch, which is a more convenient way of storing energy. The carbohydrate in a grain of wheat, for instance, is mainly starch. But starch is insoluble in water. So, when we eat a slice of bread made from wheat grains, one of the first stages in its digestion is the breaking down of the starch into sugars, which our bodies can absorb more easily. One of the most important starches in higher animals is glycogen, which is synthesized in both the liver and muscles. The liver acts as the storehouse of this starch. When active, the muscles exhaust their supplies of glycogen by breaking it down into energy-releasing sugars. These supplies are replaced by glycogen from the liver.

Cellulose, which is the most abundant organic compound on earth, is built up from long chains of sugar molecules, and often contains more than 1000 of these basic carbohydrate units. Plants use cellulose to manufacture their cell walls. Thus, it acts as a skeleton providing much of a plant's structural strength; but whereas animal skeletons are mainly inorganic, with a high percentage of calcium, phosphorus, and magnesium, plant skeletons are essentially

The diagram (right) gives approximate percentages (by weight) of basic foods and other materials in a variety of cultivated plants (see colour key below). Nitrogenous substances include proteins. Free acids—mainly citric, lactic, malic, succinic, and tartaric—are important sources of energy in citrus fruits. Ash includes valuable inorganic materials such as minerals. Note that many of these plants consist largely of water.

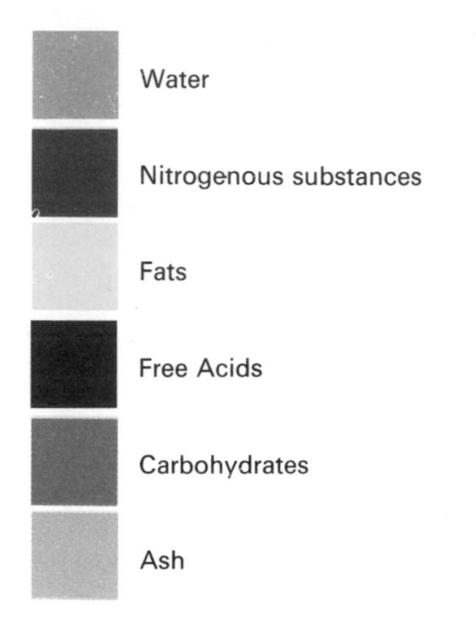

Below: starch grains of (1) potato, (2) maize, (3) rice, (4) wheat. Such grains are the form in which many plants store carbohydrates, a glucose molecule being converted into starch by the removal of one molecule of water.

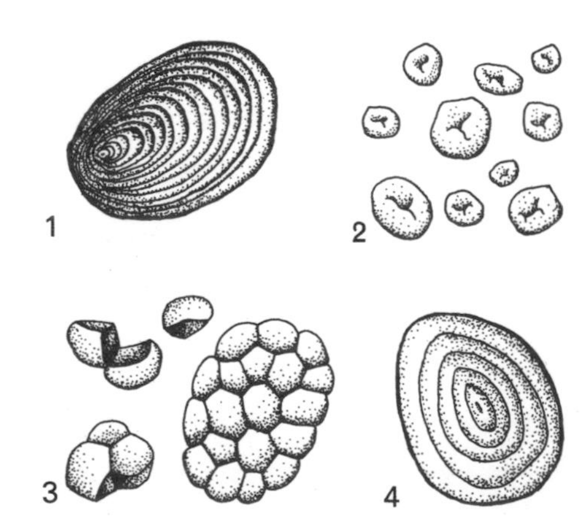

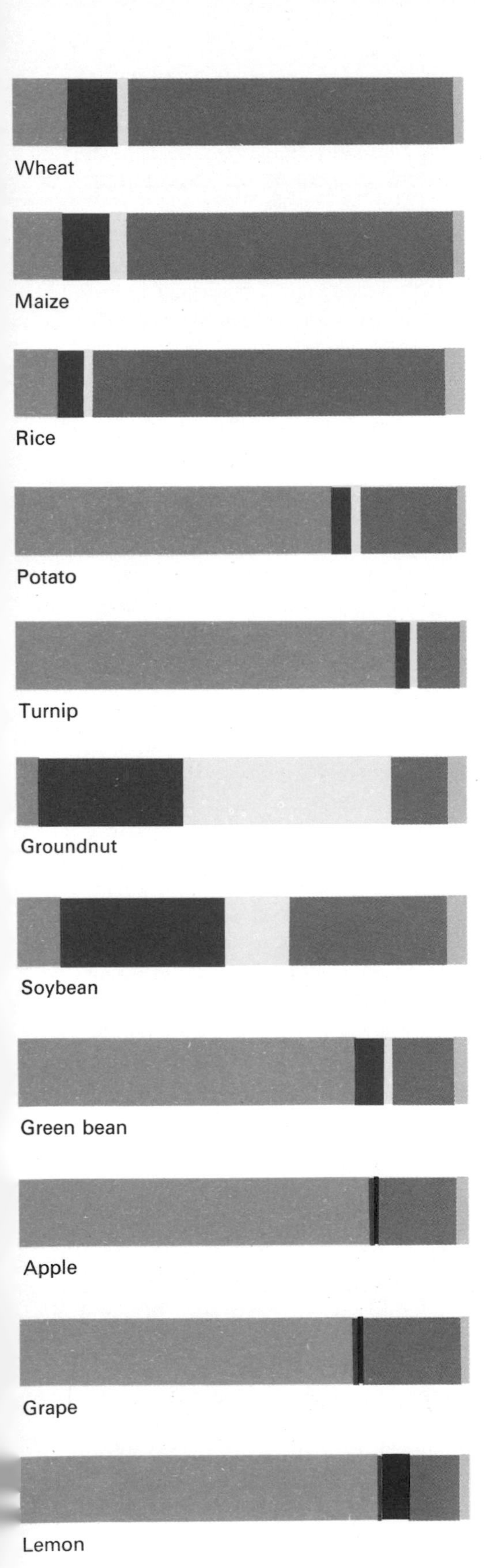

organic, consisting of carbohydrate together with small amounts of calcium, potassium, and silicon. Cellulose is indigestible to some animals, including man. But many other animals, such as cattle and rabbits, that browse on grass and similar vegetation, harbour bacteria that break down the cellulose into simpler, more-digestible carbohydrates. In woody plants the cellulose becomes *lignified*—that is, impregnated with lignin, a complex organic substance that helps to make their stems and branches rigid.

Fats and oils also contain atoms of carbon, hydrogen, and oxygen but differ from the carbohydrates in the proportions in which each element occurs. Fats are valuable as foods because they release a large amount of energy for a small volume and weight. This is partly due to the relatively low proportion of oxygen atoms in a fat molecule. Fats occur in the cytoplasm of probably every plant cell and are especially abundant in seeds, where their high energy yield is very important. Plants produce fats from carbohydrates; as their seeds germinate the fats are reconverted into carbohydrates needed for construction of the new plant tissues.

Proteins, like carbohydrates and fats, contain carbon, hydrogen, and oxygen; but they also contain nitrogen, which is essential to both plant and animal life. Plant-protein molecules are large and highly complex (some have several hundred carbon atoms and more than a thousand hydrogen atoms), and usually include small amounts of sulphur and, sometimes, of phosphorus.

Nitrogen gas accounts for about four fifths of the total volume of our atmosphere and it is often abundant in soils. Generally, however, plants are unable to take in atmospheric nitrogen, and they absorb it from the soil in the form of nitrates or ammonia compounds (the ammonia molecule consists of one atom of nitrogen and three of hydrogen). Some bacteria, however, are able to *fix* atmospheric nitrogen—that is, they are able to force nitrogen into combination with other atoms to form nitrates and other compounds. Certain of these bacteria live in the roots of Leguminosae, the family to which peas, beans, and clovers belong.

Plants form proteins when nitrates combine with carbohydrates. The first product of this combination is one of a score or more *amino acids*, which are the basic raw materials from which plants synthesize an enormous variety of proteins that are essential to the physiological needs of both plants and animals. With few exceptions, amino acids are synthesized only by plants. But with the help of these acids, derived from

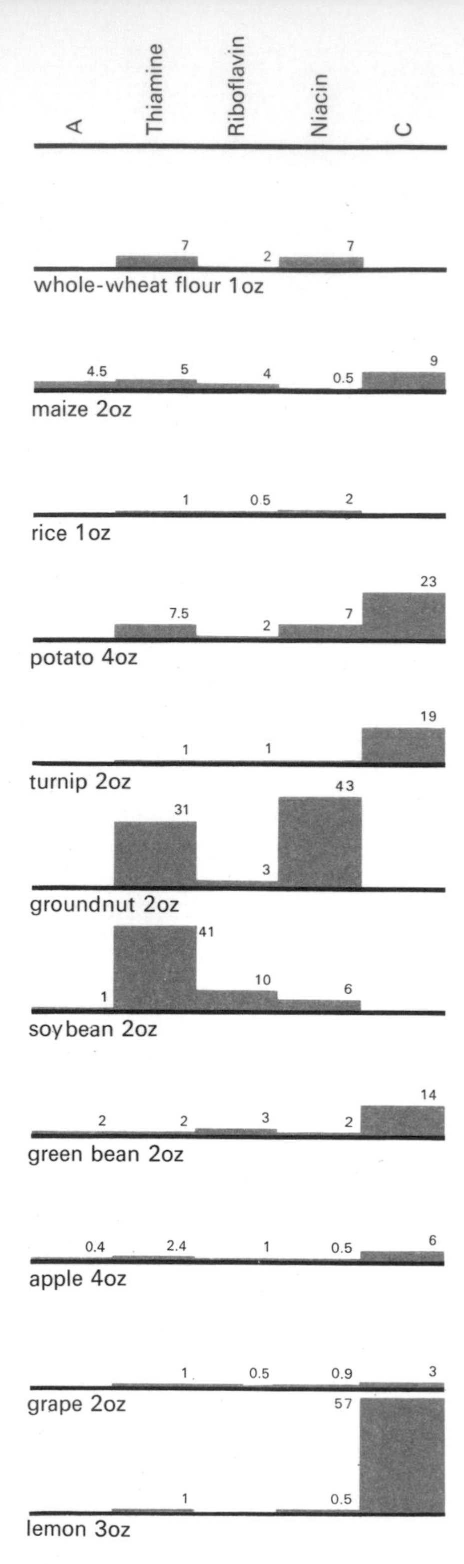

The chart shows the presence of the most important vitamins in a variety of plant foods; thiamine, riboflavin, and niacin all belong to the Vitamin B group. The figures refer to the percentage of daily requirements of a given vitamin contained in the quantity shown for each food.

plant foods, animals are able to produce many other proteins necessary to their own survival. For this reason, meat (including fish and poultry), milk, and eggs are the main sources of protein in the human diet, at least in the wealthier countries.

Minerals and vitamins, usually in very small amounts, are necessary to the health of all plants and animals. In man, for instance, minerals are important constituents of bones and teeth, of body cells concerned in the formation of blood, liver, muscles, and so on, and in the composition and stability of various body fluids. Most minerals, notably iron, calcium, and phosphorus, are taken in by plants from the soil and are thus available to animals as part of their normal food supply. In the human diet, these three minerals are often derived from plants via other animals. Our supply of calcium, for instance, comes mainly from cheese and milk; phosphorus comes mainly from cheese and liver (though it also occurs in oatmeal); and iron comes mainly from liver (though smaller concentrations are available in wholemeal bread, peas, and cabbage).

Vitamins (and other mineral salts) are essential to the enzyme systems of all organisms. In other words, they do not contribute directly to energy or growth but help to spark off the chemical reactions by which food is converted into energy or is used in the manufacture of *protoplasm*—the living matter of the cell. Plants synthesize most of the vitamins necessary for human health, and thus a balanced diet will furnish a person with all the vitamins he needs in order to keep healthy. Many of the tonic medicines used by people suffering from vitamin deficiencies contain vitamins derived from plants via animals. This is because, in some animals, the concentrations of certain vitamins are especially high.

Growth and Energy Release

All the carbohydrates, fats, and proteins necessary to a plant's life are built up, by a series of complex processes, from glucose and are then stored (with minerals and vitamins) in various parts of the plant until they are needed. But before they can be assimilated—that is, translated into the living tissues of the growing plant—they must be broken down into simpler forms. This is the process, common to plants and animals, that we call *digestion*. In almost every case, the process requires the food to take up, or be dissolved in, water, after which the breaking-down stages are carried out by groups of enzymes. Each enzyme involved is specific to a particular job of work.

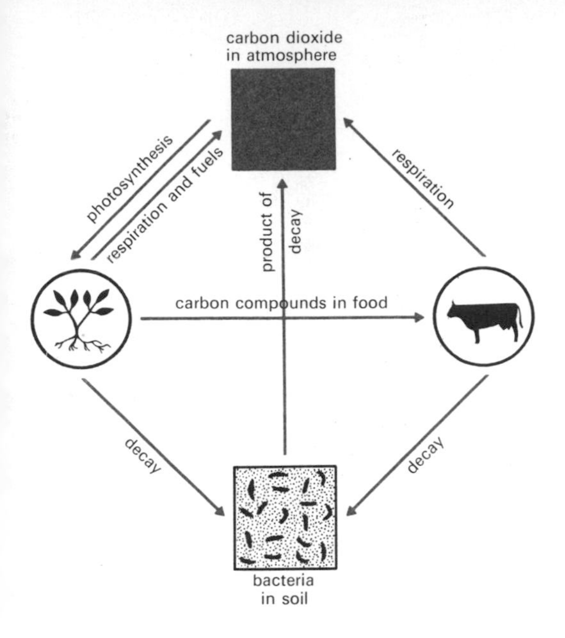

Above: the carbon cycle. Our atmosphere contains only 0.034 per cent carbon dioxide, but this is essential to the existence of plants and, hence, animals. Respiration by plants and animals, bacterial decomposition of their remains when they die, and the burning of fuels all liberate carbon dioxide and so help to maintain the supply of the gas in the atmosphere.

Below: the nitrogen cycle. Nitrogen accounts for about 80 per cent of the volume of atmospheric gases. It is also an important constituent of protein, which forms most of the dry weight of protoplasm. Animal excrement and dead plants and animals are converted by soil organisms into nitrates, the only form in which nitrogen can be absorbed by living plants. Other sources of nitrates include nitrogen-fixing bacteria, and electrical discharges in the atmosphere. Decomposition of dead organic matter also returns some nitrogen to the atmosphere.

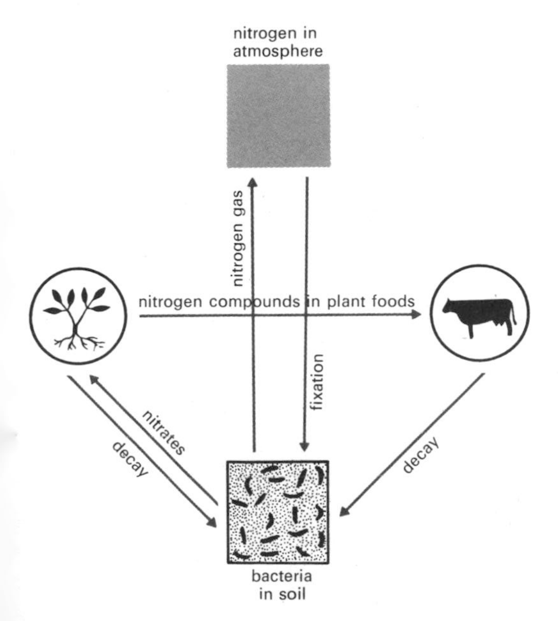

Thus, one group of enzymes is responsible for breaking down starch into sugars, another converts fats to fatty acids, another reduces proteins to amino acids.

Now, digestion, assimilation, and indeed all the manifold activities involved in a plant's every-day affairs demand a high investment of energy. We have seen that all the energy available to a plant is derived from the sunlight that triggers the photosynthetic process. This energy is locked up in the primary product, glucose, and much of it is passed on to the other carbohydrates and to the fats and proteins that the plant synthesizes from glucose. The processes by which the energy is released and made available for growth and other activities are called *respiration* and (like digestion) they occur in both plants and animals.

The respiratory processes are immensely complicated and probably involve at least 30 chemical reactions. Like digestion, they are concerned with breaking down carbohydrates, fats, and proteins into the much simpler compounds from which they are derived. There are, however, two reactions of key importance. One of these is a reversal, so to speak, of photosynthesis—the breaking down of glucose into carbon dioxide and water. Just as the energy of sunlight is needed to trigger photosynthesis, so the breaking down of glucose releases this energy. While locked in the sugar molecule, the energy is in its potential form. With the breaking down of the molecule, the released energy becomes kinetic and will run to waste unless it is used immediately. The plant uses it to promote the second key reaction.

One of the products of the assimilation of amino acids into plant and animal cells is a substance called adenosine ($C_{10}H_{13}N_5O_4$), which has a tendency to combine with phosphorus. It is the job of the kinetic energy released from the glucose molecule to bring about the combination of adenosine and phosphorus. The most important compound resulting from this reaction is adenosine triphosphate (ATP)—that is, an adenosine molecule to which a chain of three phosphate units (PO_4) is attached. The chemical bond between the adenosine and the phosphates is relatively weak. But that between each phosphate unit, and especially that between the second and third units, is immensely strong. In other words, these latter bonds represent a substantial investment of energy, which can be released to do a job of work by detaching one or more units from the ATP molecule. This molecule, then, is a miniature power pack that provides the energy needed for all the physical and chemical transformations that occur in plant and animal organisms.

Plant Physiology

We have considered some of the most important activities of plants, and we have seen that plants synthesize all the food substances necessary for human survival. One of the interesting things about man's use of plants is not simply the enormous variety of species that he cultivates but the variation, from species to species, of the particular *parts* of plants that he uses for food and other purposes. In some species, such as the turnip, it is the root that we value; in others, such as sugar cane, it is the stem; in others, such as cabbage, it is the leaf; in others, such as most nuts, it is the seed; in yet others, such as the cherry, it is the palatable tissues enclosing the seed. As we shall see in later chapters, the purpose of agriculture is not only to provide food in abundance but also to develop and improve, in each species, the particular part of the plant that is of value to man. In order to understand some of the problems of plant breeding and selection, then, we need to know a little about the role played by the root, stem, and leaf in the economy of the plant as a whole.

The typical seed falling, or deposited, on the ground responds to the problem of survival in two ways. First, it thrusts a rootlet downward in order to anchor itself to the soil; second, it pushes up shoots that eventually will develop into stems and leaves. From the initial rootlet will develop the *primary* and *secondary roots* of the growing plant—a complex fibrous system that, in many species, is more massive than the array of stems and leaves above ground.

Roots have three main roles to play in the life of a plant. First, they anchor the plant and help to keep it upright. This role is seen most strikingly, perhaps, in the large stay- and prop-roots that are partly above ground level and help to buttress the trunks of some tropical and other trees. Roots of this kind also enable plants such as the mangrove to spread laterally. The second role of the root is to store food—a role

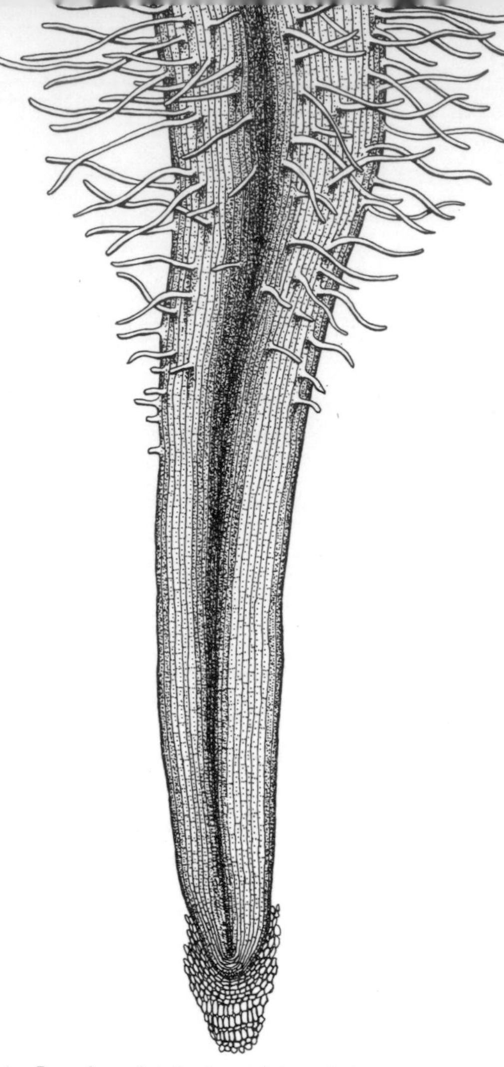

Drawing of a plant root (above) shows how the tip, which has to push its way between soil particles, is protected by a cap of cells. Above the tip is the growth zone, and above that are root hairs, which absorb water and nutrients from the soil. As the root grows, old hairs die and new ones form at a constant distance from the advancing root cap. Drawings (below) show four stages in the development of roots and stem from the seed of a castor-oil plant.

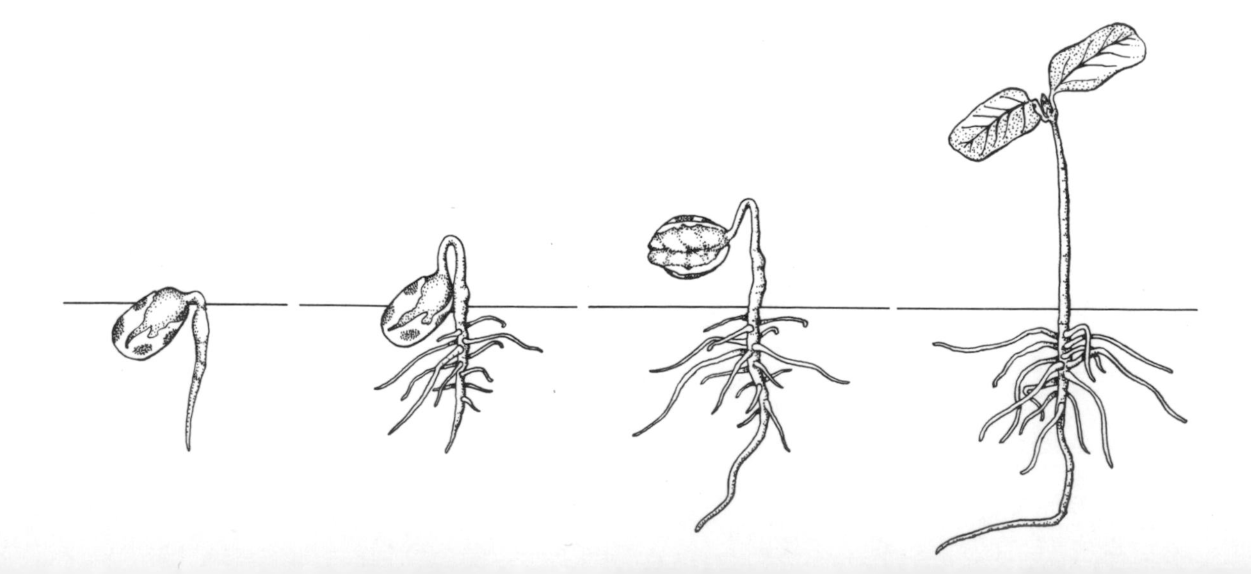

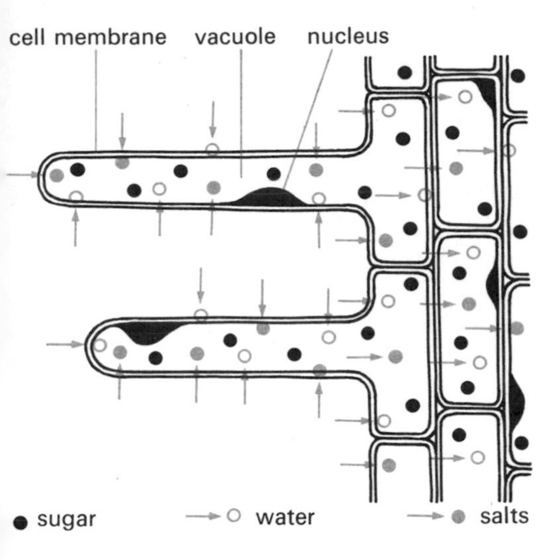

Each root hair grows from, and is an extension of, a cell on the surface of the root. Its purpose is to increase the surface area of the root and, thus, the amount of material that can be absorbed from the soil. The diagram shows the passage of nutrient salts (ochre dots) and water (ochre circles) into the root hairs, which already hold sugar (black dots) in the vacuoles. Below: the roots of different plants vary widely in appearance and size. Those of grain crops, such as barley (1), consist of thread-like seminal roots, which may later disappear, and thicker, permanent, adventitious roots that arise from the base of the stem. Roots not only absorb but also store material from the soil. The massive underground forms of the carrot (2) and turnip (3), consisting of tap-roots with smaller branching roots, are examples of such storage organs. Tubers, such as those of the dahlia (4), are not roots, but underground stem extensions. Many plants are raised from tubers instead of from seeds, the tubers developing adventitious roots soon after planting.

especially apparent in the massive roots of plants such as the carrot. The third role is to provide raw materials for growth by importing water and mineral salts from the soil.

Clearly, the absorption of food materials from the environment is vital to a plant's survival. But what, exactly, do we mean by "absorption"? Between the interior of a cell and its environment there is an apparently solid barrier consisting of the cell wall and the cytoplasm, and any exchange of substances between the cell and its environment involves the penetration of this barrier—which we call the *cell membrane*. Such penetration takes place in one of three ways. The simplest way is by *diffusion*—the mechanism by which carbon dioxide enters the chloroplasts. Imagine two solutions of water of equal volume are separated by a membrane. One solution contains six grams of salt, the other four grams. Now, if the membrane is *permeable* to (allows the passage of) salt, the latter will slowly diffuse through the membrane from the six-gram solution until the concentration of salt in both solutions is equal at five grams. The greater the difference in salt concentrations between the two solutions, the faster will be the rate of diffusion.

Diffusion occurs only with substances dissolved in solutions. But what about exchange of the solvents themselves? This takes place by a mechanism called *osmosis*, and it enables a cell at any given moment to import or export water without losing or gaining any nutrients that may be dissolved in the water.

Both diffusion and osmosis are automatic, passive mechanisms that simply equalize the concentrations of the solvents and *solutes* (dissolved substances) on either side of a cell membrane. But how do plants cope with a situation in which a needed solvent or solute is already present at a higher concentration within the cell than outside it? Conversely, how does a cell export a solvent or solute whose concentration is greater outside the cell than inside? The cell does this

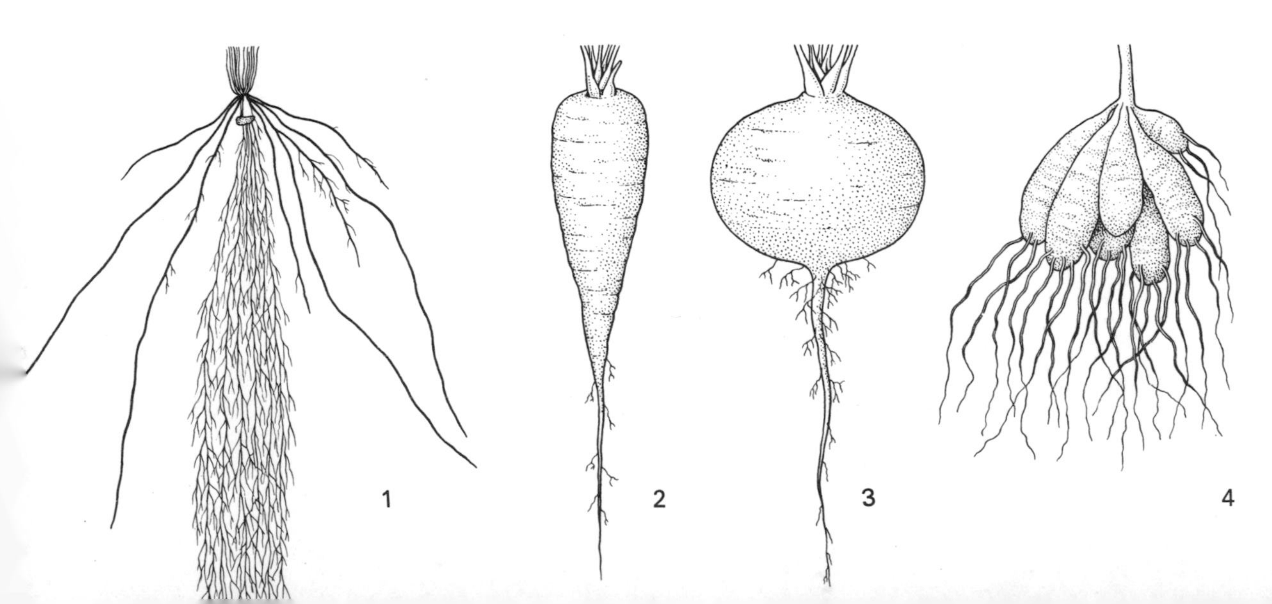

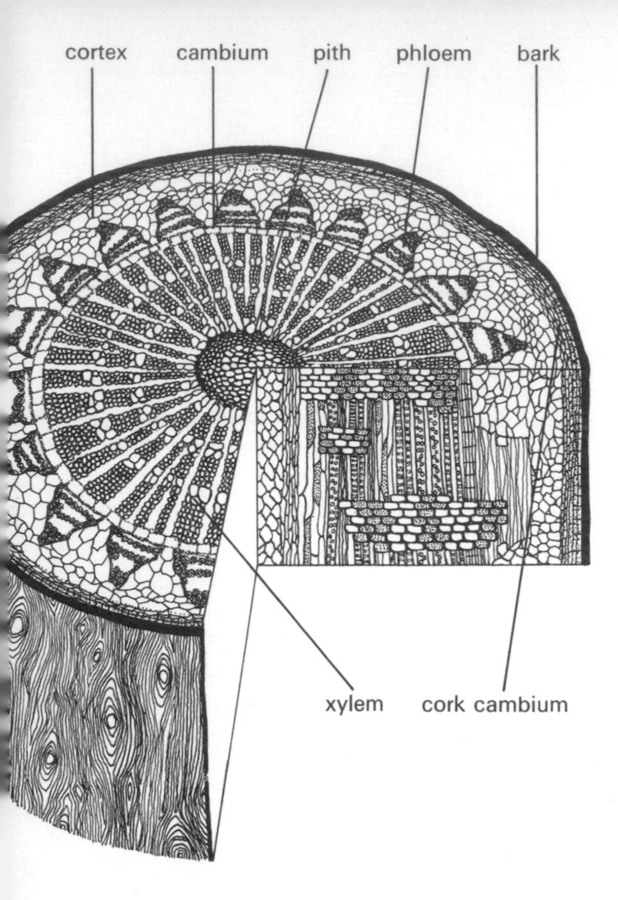

by a process called *active transport*, which is different in kind from either diffusion or osmosis because it is an active response by a plant to its needs and requires a considerable expenditure of energy. Moreover, since it involves the selection of different substances at different times, it means that the cell membrane must be able to vary its permeability to many substances—on some occasions allowing their passage, on other occasions preventing it.

A plant absorbs water and soil nutrients through hair-like filaments that develop near the tip of the root. Each of these root hairs is an outward extension of a single surface cell, sharing with it a common cytoplasm and *vacuole* (the sap cavity in the centre of the cell). Evidently the purpose of the hair is greatly to increase the surface area of the cell that is available for food absorption.

Moving upward from the root in our typical plant, we come to the *stem*. In the case of a lofty tree, such as a California redwood, the stem (which includes the trunk and branches) looks as if it is the most important part of the plant. In some smaller plants, on the other hand, the stem gives the appearance of being merely the structural link between the main centre of photosynthesis (the leaves) and the supplier of water and nutrients (the root system). The pattern of stem growth from species to species is determined by the need of the leaves to be in a position that favours the photosynthetic process. For some plants this position is near the ground; for others the leaves must be hoisted to a considerable height above the competing vegetation. The leaves' requirements are also reflected in the extent to which branching stems give lateral width to the plant. The stem is also concerned with transportation: it is responsible for conducting water and soil nutrients from the root to the leaves and for conducting foodstuffs from the leaves to the rest of the plant.

The central, conductive part of the stem is called the *vascular system* (from the Latin *vasculum*, "a small

The sectional drawing (above) shows the complex structure of a typical woody stem. The vascular cylinder, consisting of an inner portion (xylem) and an outer portion (phloem), with a thin layer of cambium cells between them, is clearly visible. Outside the cylinder are the cortex, which stores food, and the cork cambium, which develops the corky protective layer of bark. As the stem thickens with age, the bark splits and the resulting fissure is filled in from below. The splitting may take various forms—which accounts for differences in the surface texture of the bark from species to species. The examples below (from left to right) are plane tree, silver birch, and oak.

vessel"), which forms a continuation of similar systems in both root and leaves. The tissues of this system are mainly of two kinds—the *xylem*, or wood, which conducts water and mineral salts upward from the root and also provides much of the mechanical strength of the stem; and, usually outside it, the *phloem*, which conducts the synthesized foodstuffs in all directions but mainly downward from the leaf. The xylem provides a plant with its internal plumbing. The greatly elongated xylem cells die early in a plant's life, leaving a series of hollow tubes that are filled with a continuous column of water moving upward from root to leaves. The inner core of the vascular system is occupied by the *pith*, a mass of unspecialized tissue. Between the outer edge of the xylem and the inner edge of the phloem is a thin layer, or growing zone, called the *cambium*, which continuously adds cells to both xylem and phloem. The familiar annual rings in trees and other woody plants are due to seasonal differences produced by the cambium as it thickens the xylem tissues.

Enclosing the vascular system is the *cortex*, which is concerned mainly with storing food and is similar in constitution to the pith. The outer surface of the cortex is protected by a layer of corky cells produced by a cork cambium, and in trees it takes the form of bark. Its protective value lies partly in its mechanical strength and partly in its impermeability to water.

In most woody plants the vascular system appears in cross-section as a clearly defined cylinder within the cortex. The xylem forms a thick inner zone enclosing a very small area of pith. Then comes the cambium (which cannot be seen with the naked eye) and, finally, a thin outer ring of phloem. In herbaceous plants, including garden flowers, in grasses and cereals, and even in the shoots of young trees, however, the vascular system comprises a series of isolated bundles of xylem and phloem scattered about the pith, which is continuous with the cortex.

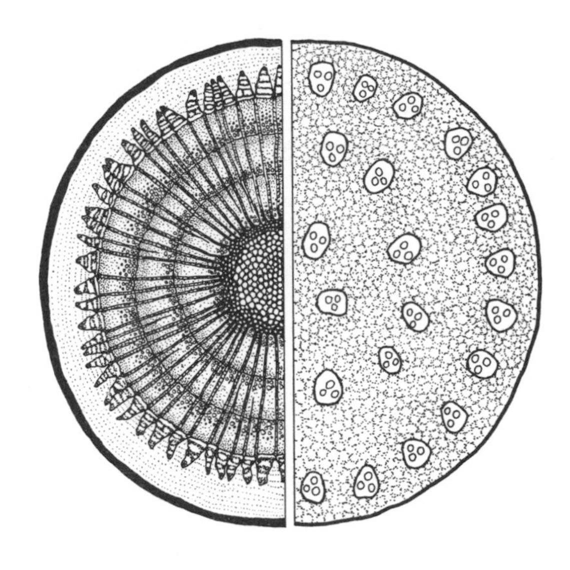

The sectional drawings (above) reveal the differences in stem structure between a woody plant (a Moojanee tree, left) and a soft-stemmed plant (maize, right). In the Moojanee, the relationship between vascular system, pith, and cortex is similar to that in the drawing on the opposite page. Note the two annual growth rings passing through the radial xylem cells. In contrast, the maize stem has a number of vascular bundles scattered about the pith, which is continuous with the cortex.

Below: plant reproduction from stems. The potato is a tuber (thickened underground stem) that, in cultivation, gives rise to a crop of other tubers (1). The strawberry plant spreads by developing surface stems, or runners (2). Solomon's seal, a member of the lily-of-the-valley family, propagates itself (as do many grasses) by means of an underground stem called a rhizome (3).

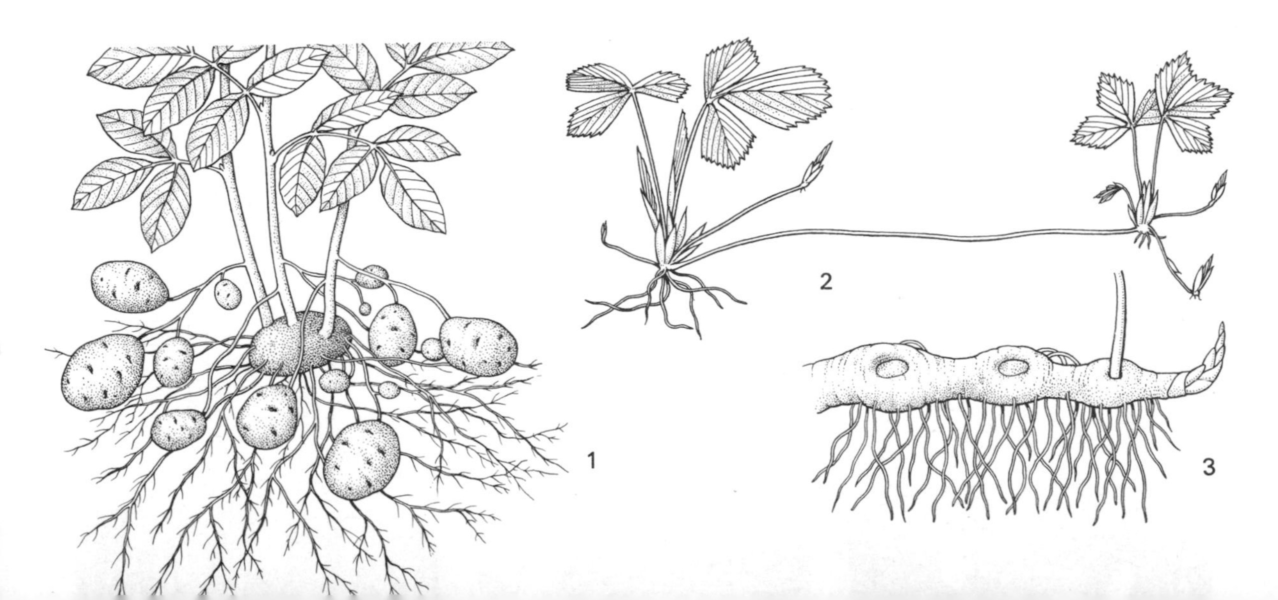

The *leaf* is attached to the stem at a point called a *node*, which is a centre of particularly active growth. Nodes do not occur at random but develop, at particular intervals and at particular points on the circumference of the stem, in a pattern that varies according to the species. Sometimes the leaf is attached directly to the stem; more often, however, it is joined to the stem by a stalk, or *petiole*, that conducts water and food between leaf and stem and, owing to its flexibility, enables the leaf to position itself most favourably to receive sunlight.

Although leaves of different plant species differ greatly in size and shape, most consist of a more-or-less flat blade intersected by a series of branching or parallel veins. These veins have two functions. First, they provide structural support for the leaf blade by acting as a skeleton. Second, they transport water and food to and from the leaf; like the petiole, with which they are connected, they contain tissues that are continuous with the vascular systems of stem and root.

Between the veins, the leaf blade commonly comprises an outer and an inner section. The outer section, called the *epidermis* (from Greek words meaning "outer skin"), provides a sheath that protects the inner section against mechanical damage and disease. Although the epidermis is only one cell deep, its outward-facing cell walls tend to be thickened and are covered by a wax-like water-proof layer called the *cuticle*.

As we have seen (page 12), although the leaf surface appears to be continuous, it in fact contains stomata that provide an entry route for carbon dioxide. The gas passes through the epidermis via the stomata to the inner section of the leaf. This section, called the *mesophyll* ("middle leaf"), is where the processes of food-making, including photosynthesis, occur. The main site of food-making activities is immediately beneath the epidermis in the *palisade layer*, the elongated cells of which are crammed with chloroplasts.

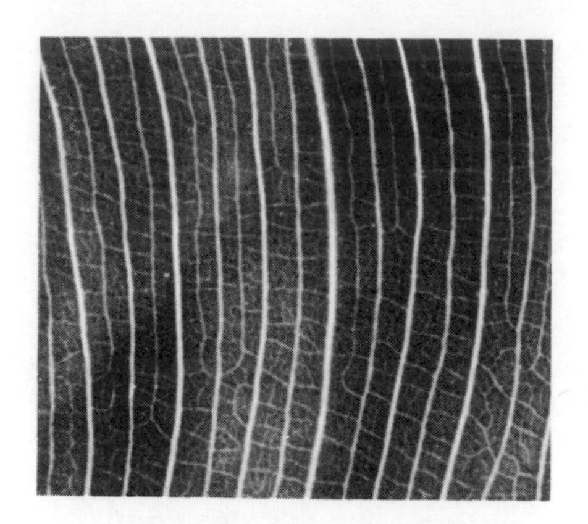

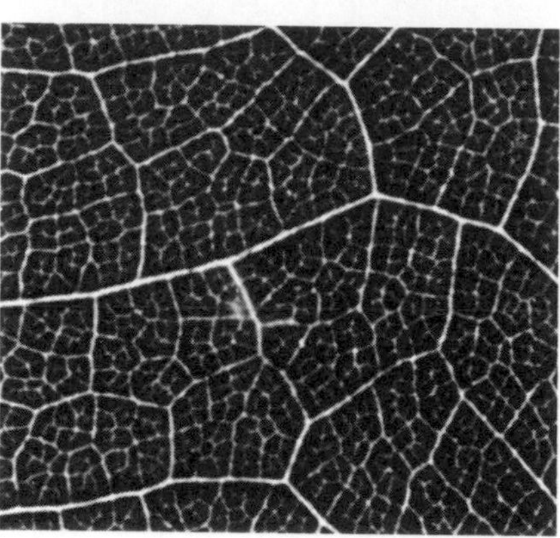

The photographs (above) illustrate the two principal ways in which veins occur in plant leaves. The upper picture shows parallel venation, characteristic of monocotyledons, in which the main veins, lying parallel with each other, are interconnected by many shorter, thinner ones. The lower picture shows netted venation, typical of dicotyledons, in which one or more prominent veins provide the frame for a complex network of smaller ones.

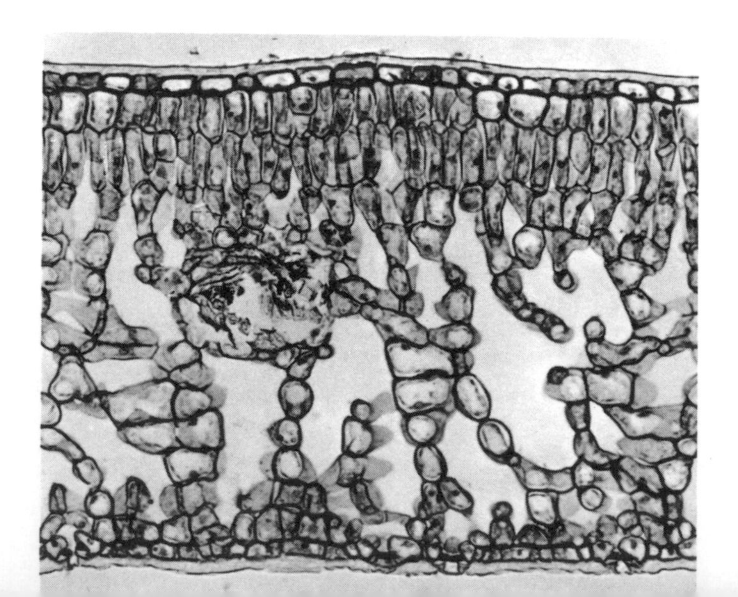

Photomicrograph (×100) of section of a holly leaf (see also upper drawing, p. 12). The upper and lower surfaces of the leaf consist of the epidermis, one cell thick, overlain (especially on the upper surface) by the waxy cuticle. The epidermis protects the inner structure, or mesophyll, which includes the palisade layer and, below it, the loose, spongy layer containing air spaces. In the centre of the picture is a section of a vein, which forms part of the vascular system.

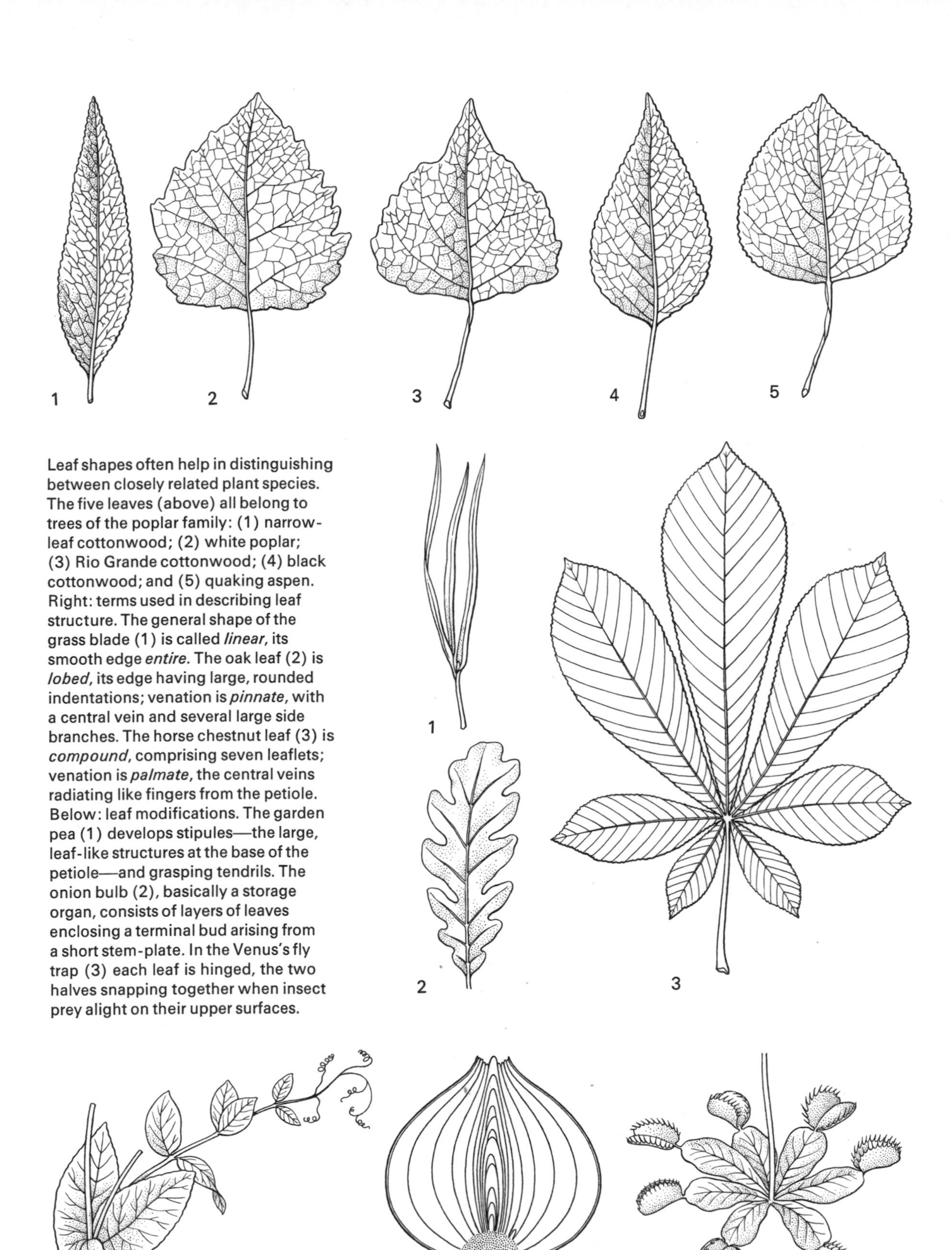

Leaf shapes often help in distinguishing between closely related plant species. The five leaves (above) all belong to trees of the poplar family: (1) narrow-leaf cottonwood; (2) white poplar; (3) Rio Grande cottonwood; (4) black cottonwood; and (5) quaking aspen. Right: terms used in describing leaf structure. The general shape of the grass blade (1) is called *linear*, its smooth edge *entire*. The oak leaf (2) is *lobed*, its edge having large, rounded indentations; venation is *pinnate*, with a central vein and several large side branches. The horse chestnut leaf (3) is *compound*, comprising seven leaflets; venation is *palmate*, the central veins radiating like fingers from the petiole. Below: leaf modifications. The garden pea (1) develops stipules—the large, leaf-like structures at the base of the petiole—and grasping tendrils. The onion bulb (2), basically a storage organ, consists of layers of leaves enclosing a terminal bud arising from a short stem-plate. In the Venus's fly trap (3) each leaf is hinged, the two halves snapping together when insect prey alight on their upper surfaces.

Beneath this layer the cells (also containing abundant chloroplasts) are much more loosely packed and are separated by air cavities, so that their structure somewhat resembles that of a sponge.

So far in this chapter, we have regarded water in plants simply as one of the raw materials in food-making or as a solvent in the digestive process. In fact, only a small proportion of the water in a plant is used in either process. Most of the remainder—which may account for nine tenths of a plant's weight—is constantly on the move, carrying raw materials and foodstuffs, via xylem and phloem, to every part of the plant. The pressure of the water has the effect of stretching each cell wall and making it rigid. Such a state is called *turgidity* and it enables many woody plants to keep erect until their cell walls become lignified and self-supporting.

Although the proportion of water weight to total

Water pressure that is greater in the xylem cells than in the soil has forced the droplets of water on to the surface of these strawberry leaves (right). This process, called guttation, occurs at points on the leaf surface where certain stomata are permanently open.

Turgidity enables soft-stemmed plants to remain upright (below, left). Denied water at its roots, a plant continues to transpire water from its leaves and so will soon wilt (below, right).

weight varies from species to species, any given species' water tolerance is strictly limited. In other words, its supply of water must always be kept within fairly narrow limits. Nutrient-rich water, as we know, is constantly being absorbed by the root. It follows that, at any given moment, an equal quantity of water must be expelled if the optimum level is to be maintained.

Plants do this by a process called *transpiration*, which occurs almost entirely in the leaf. Water enters the leaf via the xylem cells in the veins. From there, some of it passes to the mesophyll and palisade layer, where it is milked of its nutrients and is used in photosynthesis; the oxygen that forms a by-product of photosynthesis (see reaction, page 13) diffuses through these cells and enters the atmosphere via the stomata. Much of the remainder of the water from the xylem enters the spongy cells beneath the palisade layer. Some of this water is retained; the rest evaporates through the cell walls and enters the atmosphere (again via the stomata) as water vapour. Many plants expel their own weight of water every day by this means, and a large tree may transpire as much as 250 gallons daily.

The rate of transpiration depends—rather as the rate of perspiration depends in man—partly on the state of the environment (temperature, humidity, and so on) and partly on the organism's internal needs. In other words, transpiration is neither directly a cause nor directly a consequence of water absorption at the root, though the two functions must obviously influence each other; moreover, transpiration is directly responsible for the upward movement of water in the xylem. Different plant species differ greatly in the amount of water they need: some flourish only in the tropical rain-forest; others can survive in semi-desert. This seems to imply that those plants whose "turnover" of water is relatively small are likely to suffer from a shortage of water-borne soil nutrients. In fact, this is not so. The various methods by which plants absorb solvents and solutes are, as we have seen, highly selective: at any moment, the necessary concentrations of nutrients can be imported into the root even if the amount of water available to carry them is small.

In this chapter we have considered the growth and functions of root, stem, and leaf only in the most general way. The details vary—sometimes greatly—between species. And, as we shall see in later chapters, it often happens that these specific differences, these particular developments of root, stem, and leaf, are the features that give a plant its importance to man.

2 The Development of Agriculture

Man has existed on this planet for probably a little more than a million years. For all but a tiny fraction of this time he has been a hunter of animals and a gatherer of roots, berries, wild grasses, and other plant foods. The name we give to this long period of man's infancy—the Palaeolithic or Old Stone Age—refers to the primitive stone implements he fashioned for use as axes, knives, and hammers. We can get some idea of the lives of the later Stone Age people from the isolated pockets of hunter-gatherers that still exist in various parts of the world. Both the Australian aborigine and the Bushman of the Kalahari Desert in South Africa lead lives that can have changed little for several tens of thousands of years. Their entire cultures, like those of their Palaeolithic forebears, turn on the single question: where to find the next meal. Their material possessions are minimal: they need little or no clothing; they build no houses, for they are constantly on the move; their domestic equipment is confined mainly to digging sticks and dishes or gourds for carrying water. The aborigine carries a stone-tipped spear and a spear thrower; the Bushman is armed with a bow and poisoned arrows. Both have developed the arts of tracking and hunting to a point hardly approached elsewhere in the world. Yet, to most people living today, their lives would seem intolerably harsh and limited in scope.

We may never know exactly when or in which regions of the world our earliest recognizably human ancestors first arose. It is plain that, by the late Palaeolithic, the nomadic existence forced upon the hunter had carried man into most of the habitable parts of the world. He was bound to keep on the move—following animal herds and searching for plant foods in due season—until he had discovered the secret of raising plants for his own use.

It is almost certain that this secret was accidentally discovered about 10,000 years ago—at the beginning of the Neolithic, or New Stone Age—in the temperate grasslands of the Near and Middle East that extended

Hundreds of thousands of years separate man's beginnings and his first attempts at cultivation. Throughout this period, man was a hunter of wild animals (such as the deer, right, painted by late Palaeolithic peoples on the walls of a cave at Albocácer, Spain). The name Palaeolithic, or Old Stone Age, refers to the tools made by such hunters, of which the hand axe above is an example.

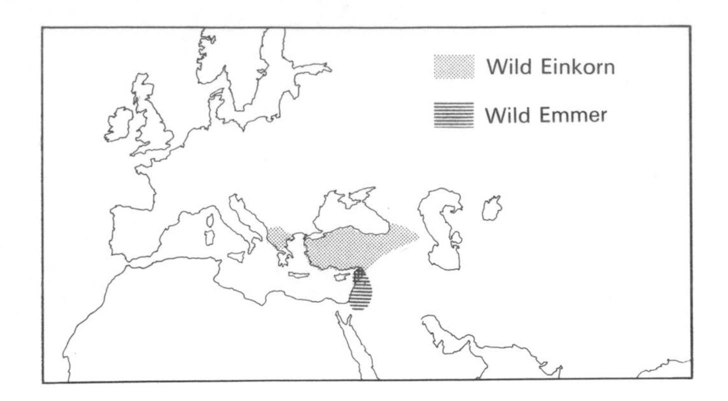

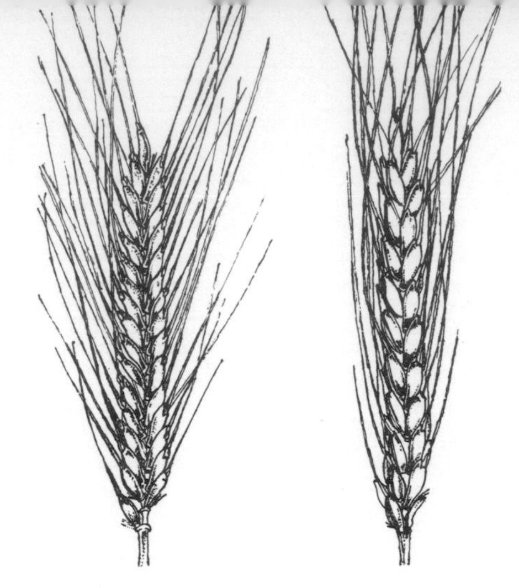

from the northern coast of Africa through much of Palestine, Syria, Iraq, and Iran. For it was in this "fertile crescent" that man began his earliest, hit-or-miss attempts at farming. We can only guess exactly how the hunters first grasped the mystery of plant growth. We know that toward the end of the Palaeolithic many hunters had sickles made of chips of stone set into wooden or bone handles, and this suggests that foods such as wild grasses or grains were abundant enough to require the aid of a tool for cutting. Large quantities of food raise the question of storage. It is possible that the harvesters (who were almost certainly women: the men were animal hunters) noticed in the spring that grain seeds spilt around their primitive granaries at the previous harvest time had, as if by magic, begun to sprout from the soil. Later in the year, it was seen that these young plants had matured into the same seed-bearing adult plants that the people regularly harvested in the autumn. Once the link between seed and food-bearing adult plant had been grasped by the gatherers, the purposive planting of seed was inevitable and the Neolithic revolution took its first giant step forward.

The maps (left, above and below) show the distribution in the Middle East of the wild ancestors (drawings) of modern wheats and barleys, which were probably the first crops man learnt to cultivate. The development of agriculture is justly called the Neolithic revolution. It not only provided a radically new source of food; it also changed man's way of life from nomadism to permanent settlement, which is a prerequisite of civilization. The centre of the revolution was the Middle East. At first the peoples of this area merely harvested the wild grasses; later they discovered how to raise them from seed and how to improve their quality and yield by selection. In time, the revolution spread to other regions, where different wild grasses (notably rice in eastern Asia and maize in the Americas) were brought into cultivation. Probably every plant food important to us today had been domesticated by the end of the Neolithic.

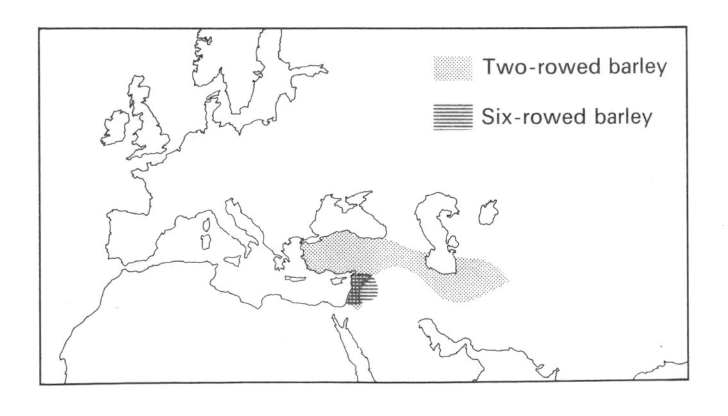

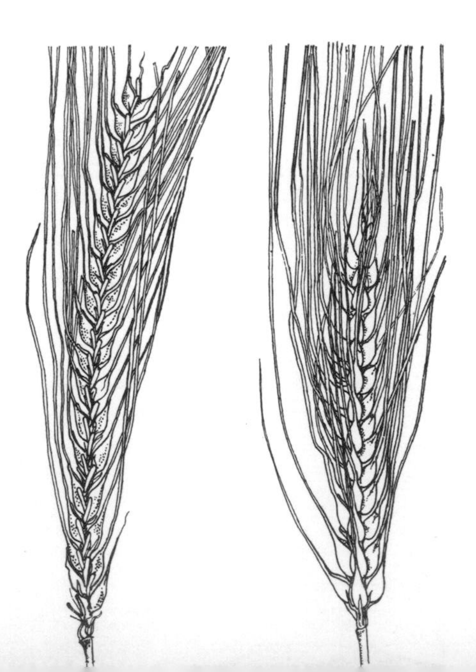

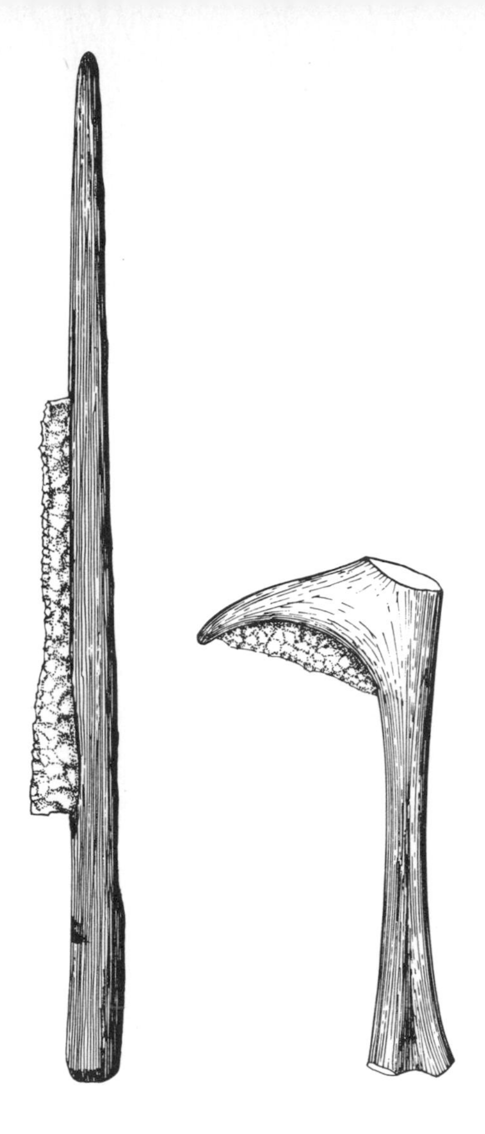

Apart from grasses and berries, the staple plant foods of the hunter-gatherers consisted of roots and bulbs. While their menfolk were hunting, the women sought out these plants with the aid of simple digging sticks. Doubtless the tribespeople learnt to remember places where such roots grew abundantly, and would return to them in later years. It seems likely that the women eventually noticed that plants of all kinds grew better and yielded a higher quantity of food in those patches of soil that had been disturbed in previous seasons by their digging sticks. In this way, perhaps, they learnt the value of *tillage*, or soil cultivation. The digging stick was the crude but effective forerunner of the hoe and plough.

About this time, the relationship between man and food animals was beginning to change. In times of plenty, instead of killing all the animals they could catch, the hunters would spare some of the healthier ones and would herd them on their wanderings in search of plants. In this way cattle, sheep, goats, horses, asses, camels, buffaloes, and other animals were domesticated. Herding had important implications. It not only eliminated the hard slog of hunting day in, day out; it also guaranteed the careful herdsmen a permanent supply of meat in good times and bad and, given good pasture land, enabled them to call a halt, at least temporarily, to their ceaseless journeyings from place to place. For those herdspeople who were beginning to understand how to domesticate plants as well, an entirely new way of life began to unfold.

The fertile crescent provided grasslands that not only were good for raising crops but supported enough wild animals to form the nucleus of herds. Agriculture (from Latin words *agri cultura*, meaning "cultivation of the land") has from earliest times included the tending of livestock as well as the raising of crops. After harvesting, crop fields were used as

Above: Neolithic Egyptian sickle (left) with flint blades in wooden handle, and (right) an improved version from Denmark with flint set in bone. Below: harvesting grain in Egypt, about 1100 B.C. The man cuts a handful of ears at a time, the woman places them in her basket.

grazing land for the animals, and their dung and urine returned to the soil some of the phosphorus, potassium, nitrogen, and other essential nutrients that the plants had drawn from it. Once again, man gradually learnt a lesson from this and began to enrich his land by systematically digging manure into the soil before planting seed. (The verb "manure," incidentally, has the same derivation as "manoeuvre": both mean, literally, to work by hand.)

All these discoveries—the knowledge of the basic essentials of planting, tillage, animal herding (and, later, breeding), and manuring—did not come overnight. They were made, at different times and in different places, over a period of many thousands of years. At first they could be applied only in the temperate grassland regions of the fertile crescent and elsewhere. As we have seen, the desert still denies the aborigine and the Bushman the opportunity to break out of their Palaeolithic cul-de-sac; so, too, do dense jungles and forests in other parts of the world. Historically, agriculture has always been imported into tropical or temperate forest regions by peoples who have learnt the arts of farming in grasslands.

Yet, however slow it was in coming, however fortuitous the discovery of its secrets, agriculture remains the most important cultural development in human history. It enabled man, for the first time, to bend nature to his will, to plan his life, and, occasionally, to direct his energies toward something other than a ceaseless search for food. But the most decisive

With agriculture came the need for tools to cultivate the soil. The simple hoe (above) was cousin to the immemorial digging stick. The ox plough (below), exploiting greater muscle-power than a man's, enabled each farmer to cultivate a larger acreage than ever before. (The wooden models on these two pages are Egyptian and are about 4000 years old.)

benefits of agriculture came, at first, only to the more successful farmers with large healthy herds and highly productive crop lands. It sometimes happened that these people, after a few good seasons, discovered that they had a surplus of food over and above their foreseeable needs. It is from agricultural surplus, not agriculture as such, that civilization flowed in all its richness and complexity. For it enabled the farming communities to import, at first modestly but later in abundance, a variety of commodities, techniques, and ideas in exchange for food. These imports varied greatly from region to region, according to need and availability. But the commodities no doubt included new farming tools, cooking utensils, building materials, weapons, and improved animal stocks and plant strains. Among tools, the most important was the plough, which was in use in Egypt, Iran, and India at least 5000 years ago. It not only expanded the scale of agriculture from small-plot cultivation to the farming of large fields; it also transformed the ox and other beasts (hitherto valued only as food) into draught animals. Soon it was realized that animals could replace human muscles in a whole range of activities. The ass (originally a native of East Africa) was in use as a pack animal at least 5500 years ago; and wheeled vehicles were common throughout the Near and Middle East some 1500 years later.

Greater mobility enabled the farming communities to trade over ever-greater distances. As they grew wealthier, they developed, or obtained in exchange for other commodities, increasingly sophisticated goods and ideas—luxuries such as spices and jewellery, and new methods of processing foodstuffs. Ultimately, too, they developed or acquired the two most revolutionary techniques of all—the priceless knowledge of writing and the secret of refining metals.

In the space of perhaps 5000 years, communities in the great valleys of the Nile, Tigris, Euphrates, and Indus rivers developed from small tribal farms into complex, highly organized urban societies in which farming was only one of many specialized trades or professions. It may be true that urban society is the ultimate destiny of even the most primitive peoples, but a special factor accelerated the process for the wealthier peoples of the fertile crescent. Their farming lands were concentrated in the flood plains of the four great rivers, and once or twice every year, and sometimes oftener, the rivers burst their banks and spread silt, rich with mineral nutrients, over the crop fields. Thus the farmers were able to raise heavy crops year after year without exhausting the soil.

In the model of a granary (above) an official directs two labourers (not visible) who are carrying baskets of grain, while a scribe makes a record of the stock. The figure below is kneading dough. The Egyptians were probably the first to discover that light, "raised" loaves could be made by allowing wheat dough to ferment and form gas pockets.

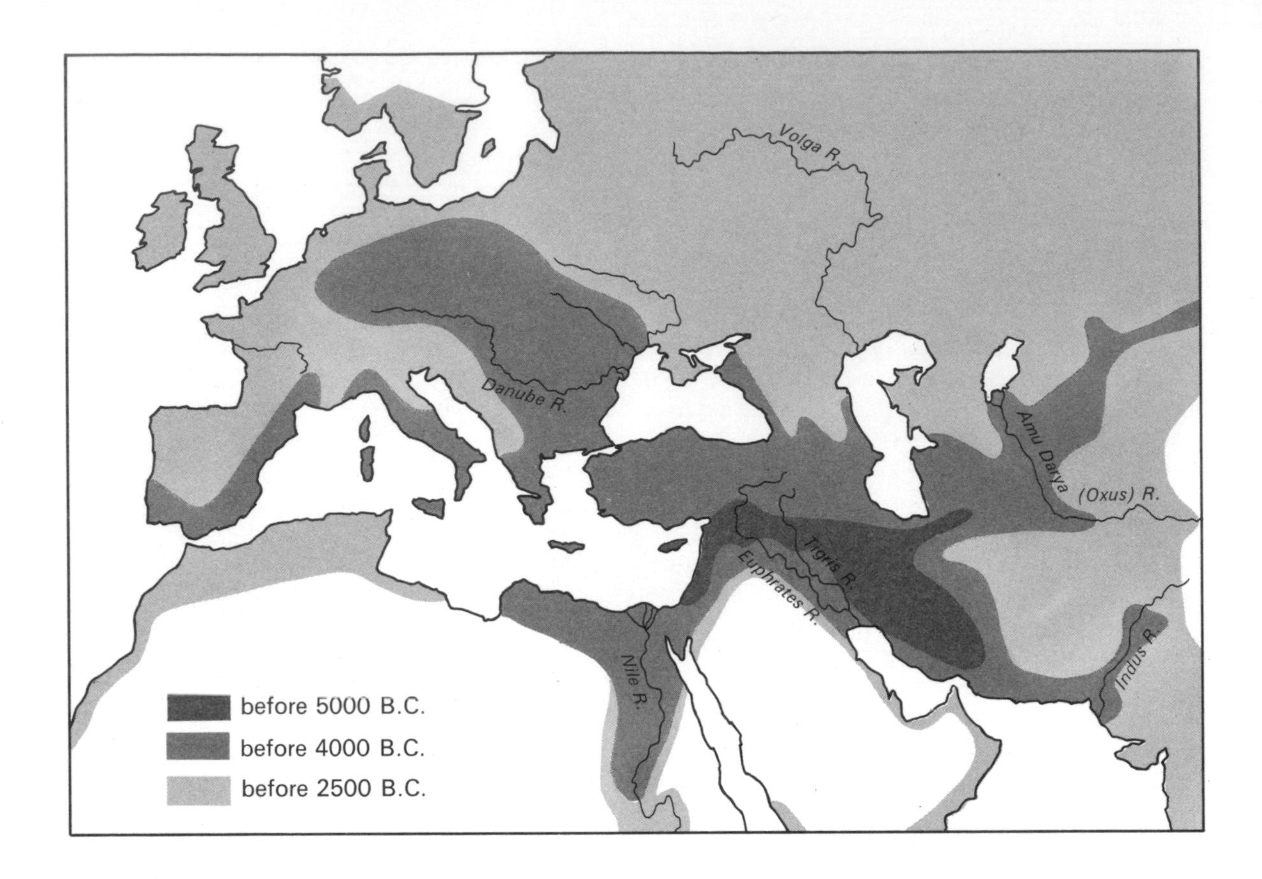

Map above shows the spread of Neolithic culture, from its beginnings (in the Middle East) to the dawn of the Bronze Age (2500–2000 B.C.). The term Neolithic refers to a stage of cultural development, not to a historical period. Many of the indigenous peoples of Europe retained their Palaeolithic hunter-gatherer cultures for several thousand years after Neolithic settlers had begun to farm the Danube valley; while north-western Europe's Neolithic cultures were contemporary with the Bronze Age urban civilizations of Mesopotamia, Egypt, and the Indus valley.

Most Neolithic communities, however, were not fortunate enough to live on the flood plains. And since they had little idea of how to regenerate soil by methods such as crop rotation, they discovered that, sooner or later, their lands became impoverished and would support neither plants nor animals. Thus, they had to move from their settlements and search for new lands to farm. From the borders of the fertile crescent, the farmers spread out in all directions from about 5000 B.C. onward, and archaeological finds have helped us to trace several of their courses. Some spread westward along the coast of North Africa. Others migrated along the northern coast of the Mediterranean, keeping south of the mountain ranges of the Balkans, the Alps, and the Pyrenees. A third wave went farther north, up the Danube valley, and eventually spread right across central and western Europe.

These migrations took several thousand years, for the tribes were not constantly on the move. As soon as they came upon a tract of good land they settled down, raising wheat or barley until the soil was exhausted, and then moved onward, driving their herds before them. As they moved westward and northward, the natural vegetation changed. On the northern Mediter-

Opposite page: food crops decorated many coins during the classical period. The barley ear (top) is on a silver Greek stater, about 520 B.C.; the grapes (centre) on a bronze Thessalonian coin, about 200 B.C.; the palm (bottom) on a silver Carthaginian four-drachma piece, 405 B.C.

ranean coast were evergreen forests; to the north of the mountains, central Europe was covered by vast tracts of deciduous forests, which, on their northern margin, gave way to colder regions of coniferous forests. The same problem of forests faced pioneer farmers in other regions, notably in Africa and in North and Central America. Indeed the history of agriculture, from Neolithic times right up to the days of the European pioneers in North and South America, South Africa, Australia, and New Zealand, has been intimately bound up with the problem of forest clearance.

This pattern of *shifting cultivation* is still common in central Africa and in southern Asia. It is a prodigal method: for every acre under cultivation it requires another 20 at least to lie idle. Many of the Neolithic tribes that introduced farming to Europe took possession of large areas of woodland and then cleared and cultivated small patches at a time. When the soil of one patch—perhaps four to six acres—was exhausted, they moved to another. When all the patches had been farmed they returned to the original one, which now, 40 or 50 years later, had had time to revert to forest, its soil enriched by the annual fall of leaves.

In many cases, however, the forests never managed to recolonize the cleared land, for goats, sheep, cattle, and other domesticated or wild animals would bite back the young shoots as soon as they appeared above ground. Under these circumstances, the thin topsoil would become exposed and would be swiftly removed by wind or heavy rainstorms. Many of the present-day "dust-bowls" around the Mediterranean are the consequence of shifting cultivation followed by overgrazing; they may also explain why so much of the old fertile crescent is now scrubland and desert.

Yet, intensive cultivation of the soil has not always brought disaster in its train. As we have seen, the silt-enriched soil of the flood plains in the fertile crescent enabled the farmers to provide enough food to sustain large populations of city dwellers. During early stages in the development of these riverside settlements, the farmers were content to allow the seasonal flood-waters to spread out haphazardly over the crop lands and then to sow their corn seeds when the waters had receded. So long as the people could retreat to higher ground when the river spilled over its banks, flooding was an effective way of bringing water and nutrients to the land. But as these communities thrived and their growing populations began to concentrate in urban centres, it became apparent that uncontrolled flooding was a wasteful as well as a dangerous method of using

The six scenes (above), from a Theban tomb of about 1450 B.C., show the gathering in of the grain harvest. In the first scene the workers cut the grain with sickles; other workers take the grain baskets to the threshing floor, where it is spread out and trampled by oxen to separate the grain seeds from the husks. Then the grain is winnowed in shallow trays, the husks blowing away in the wind, before being stored under the watchful eye of a foreman, who gives details of the crop to the scribes.

The paintings below are from another Theban tomb of about the same date. In the upper register, or row, of the picture at left the foreman leaning on his staff watches labourers gathering the grape harvest. In the middle register two men are filling and sealing the wine jars; to their right a man with a brazier stands before a statue of the cobra goddess; behind her two men tread grapes. Bottom register: the ship is laden with wine jars; to its right, rope-makers work between clumps of papyrus, their raw material. In the picture at right, the owner of the tomb sits before his food stores and watches his labourers at work.

The shaduf (above), one of the earliest mechanical irrigation methods, consists of a pole balanced on a support, with a bucket suspended from one end and a counter-weight at the other. The labourer is pouring water into a trough leading to the base of the tree. Irrigation made possible not only the raising of basic food crops but also such luxuries as this ornamental lake (below) on a private estate. (From a painting on the wall of an Egyptian tomb, about 1200 B.C.)

the water. Thus it was here, along the banks of the Nile, Tigris, Euphrates, and Indus, that farmers developed the technique of irrigation, whereby flood-waters are channelled by way of canals, dikes, and sluice-gates to every corner of cultivable land. It is likely that irrigation was first used in Egypt, where the flood plain, extending three or four miles to either side of the Nile, is more than 550 miles long. It is certain that Egyptian farmers were using dams for irrigation about 5000 years ago. Since then, the technique has spread to many parts of the world, notably to the rice-growing peoples of eastern and south-eastern Asia. Rice has to grow under water at one stage in its cultivation, and irrigation was vital to the development of the teeming communities that arose alongside the Irrawaddy, Me-kong, Red, Si Kiang, Yangtze Kiang, and other great rivers of the east.

Mixed and Specialized Farms

From the earliest days of agriculture until about 200 years ago, every farm was concerned with providing a community—the farmer's family and dependants, or a tribal group or village—with all its food requirements throughout the year. Indeed, its products went beyond food, for the farmlands were expected to furnish fuel for fires and timber for buildings, fences, and vehicles. The farmer's range of food products was enormous: it included grain and root crops, fruits, cattle, sheep, and poultry, and he had to plan his production with enormous care to avoid exhausting the precious soil on which the lives of the community depended.

This kind of self-sufficient, mixed farming has some advantages but many drawbacks. The farmer is independent: he grows all his own food, fuel, and building materials; he can even provide the raw materials for most of his clothing. But, no matter how carefully he plans, the multiplicity of his roles creates unavoidable crises: he is certain to grow too much of some commodities, too little of others. In one year he may have too much corn for the people and too little hay for the animals; in another, he finds himself short of milking cows but with too many sheep. Soil and climate may conspire against him: they may favour wheat crops, which yield high profits; yet, in catering to the needs of the community, the farmer may be obliged to use many of his wheat fields for cattle pasture or woodland—neither of which may provide him with much income.

In spite of these difficulties, mixed farming is still

Economic factors have created a trend toward large farms that specialize in a single crop. The prairies (as here in Manitoba, Canada) are particularly suited to large-scale wheat cultivation.

practised in northern Europe and in parts of North America. The farmers provide themselves with most of their daily needs and sell their surplus in the local market. Many such farms are quite small yet have endured for centuries; in Iceland, for example, some farms still bear the names given them by Norse settlers a thousand years ago.

But economic factors, as well as climate and soil, have changed the pattern of agriculture in most other parts of the world, especially in the technically advanced countries. Today, most large farms specialize in one or two crops, or else raise livestock and nothing else. This method of farming requires a high investment of capital—a great deal more than is available to the peasant farmer—since it is worthwhile only on a large scale. But it offers correspondingly greater rewards. If one is concerned solely with raising a single crop, for instance, it is possible to guarantee a certain measure of success by first of all choosing to farm in an area where climate and soil are favourable and where there is a ready supply of skilled labour. As communications develop, so more markets become available. For example, if the Australian sugar crop exceeds the local demand, the surplus can be profitably exported to Europe. If too much grain is harvested in America, it may be possible to ship part of the crop to help feed the undernourished peoples of Asia (always assuming that these people are prepared to eat an unfamiliar food).

In a country like England, most of the more prosperous farms are specialized. But even in this quite small area the climate is variable enough to encourage different specializations. Most of the grains, for instance, are grown in the comparatively dry and sunny south-east. In the west, where it is too wet and cloudy for grains, most farmers specialize in livestock because pasture grasses grow all the year round. Much of the fruit is grown in two counties, Worcestershire and Kent, where the summers tend to be warm and sunny.

On the international scale, however, we find that some countries specialize in particular crops because climate and soil, aided by traditional or scientific know-how, have enabled them to produce commodities that are either cheaper or better than their competitors. Burmese rice from the Irrawaddy flood plains, cotton from the southern United States and Egypt, rubber from Malaysia, wheat from the Canadian prairies, meat from Argentina, wines from France, West Germany, Italy, and Spain, whisky from Scotland—all these products have given their countries a major share in world trade.

Climate and Soil

The success of these specializations depends to a large extent, as we have said, on climate and soil. These two factors, indeed, determine both the type and the scale of farming in all the agricultural regions of the world. The map shows the distribution of the world's natural vegetation. Farming, at least on a large scale, is carried out only in the grasslands and in those regions of coniferous, deciduous, and tropical rain-forests that have been cleared by man for cultivation. The map is a series of generalizations. It cannot show, for instance, the effects of altitude,

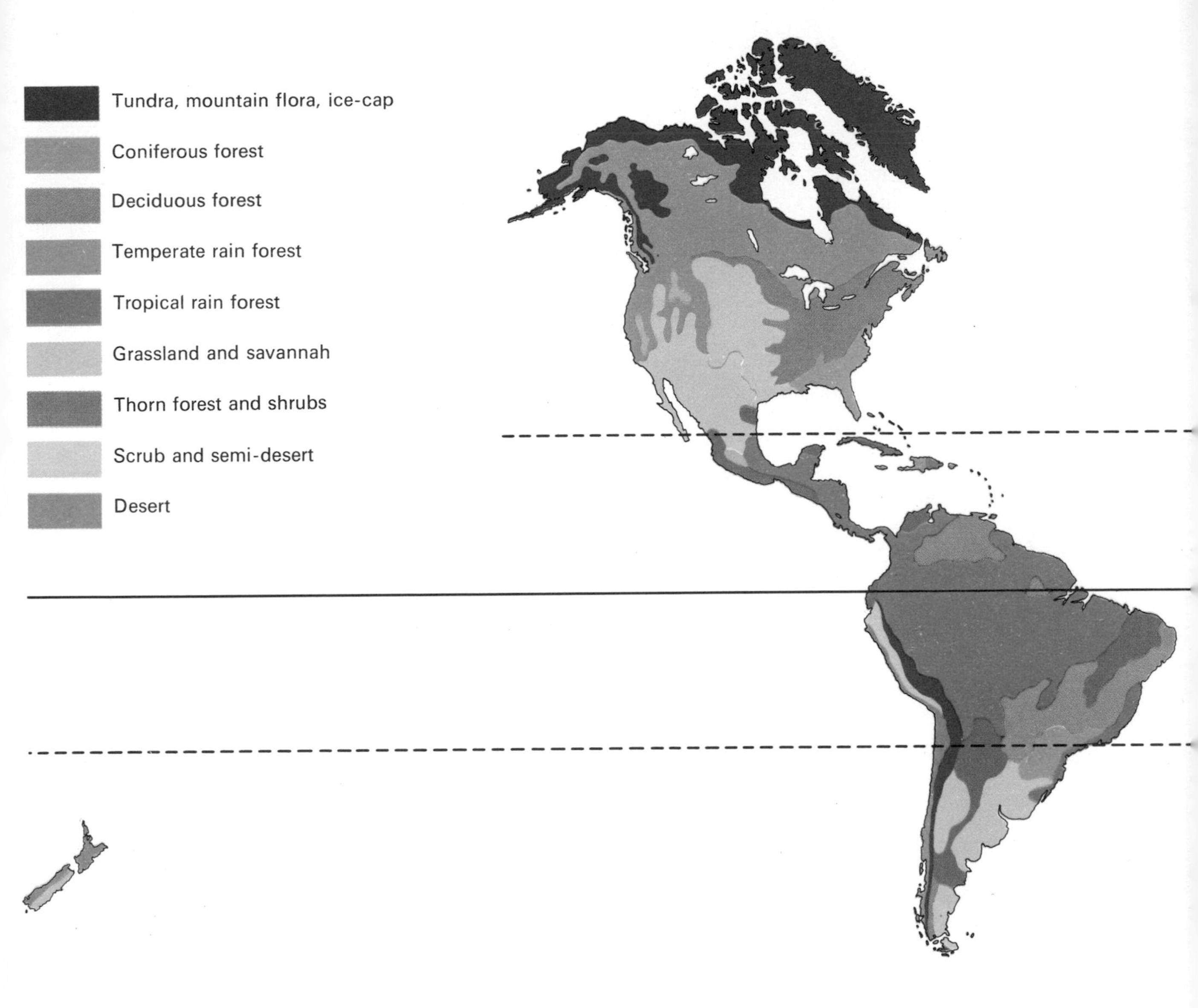

World map (above) shows distribution of main types of climatic climax vegetation. It does not show man's alterations to the vegetation picture, which in densely populated regions are considerable. The boundary between two types of vegetation is, of course, much less abrupt than is implied on the map. (See also the world soil map on pages 42 and 43.)

Diagram (left) shows in simplified form how climax vegetation growing on a mountain side changes with altitude.

which can transform the picture at a local level. The vegetation on the side of a mountain range in the tropics may run the gamut from rain-forest at its base to mosses and lichens beyond the timber line near the summit.

More important, the map cannot show the enormous variations, both in individual species and in communities of species, that occur within each vegetation zone. The types of vegetation listed on the map are what are called *climatic-climax vegetation.* "Climatic" refers to the obvious fact that climate exerts a crucial influence on the range of species that can thrive in any region. "Climax" implies some cycle of events that culminates in a particular type of vegetation; and this is precisely what happens.

Forests of massive trees cannot grow on bare rock or even on shallow soil because their roots could not penetrate deep enough to provide anchorage and because the rocky material contains little or none of the organic material that the trees need if they are to thrive. Soil is formed when the rock surface is broken down by *weathering*—a process carried out by ice and frost, by naturally occurring chemicals, and by lowly plant organisms such as algae or lichens. Soil is commonly a mixture of clay, silt, and sand particles, together with any nutrient minerals contained in the parent rock below. We call this part of the soil the *inorganic fraction*. *Humus*, which is the *organic fraction*, consists of the remains of dead plants or animals that have been broken down by soil bacteria into forms that can be assimilated by growing plants. Humus establishes a definite chemical link with clay particles, and the resulting *clay-humus complex*, to which nutrient minerals become attached, is the main centre of fertility in any given soil.

The cycle that culminates in climax vegetation often begins with lichens colonizing the smooth, hard surface of rocks. They loosen particles of the rock and, when they die, bequeath fragments of organic material that offer nourishment to more complex plants. In temperate regions the lichens, having made their small contribution to the formation of soil, are often ousted by mosses; the cycle may continue with successive dominance by tussocky grasses, brambles and other shrubs, and small birch and ash trees. Finally, the climax is reached when the soil is invaded by deep-rooting giants such as oak and beech. This vegetation represents the point at which a given plant community in the temperate zone reaches a state of equilibrium that will endure indefinitely unless it is disturbed by natural agency or by the hand of man.

The cycle can be arrested or disrupted entirely in many ways. Quite small adjustments in climate are one way. Geological factors, such as landslides or volcanic eruptions, are another. The clearing of forests for agriculture is yet another. It is a matter for debate whether the extensive grasslands of the world can properly be called climax vegetation. Clearly, the North American prairies, the South American pampas, the Eurasian steppes, and the temperate grasslands of South Africa, Australia, and New Zealand are due, at least in part, to destruction of the forests by fires started by lightning or man, and to subsequent grazing

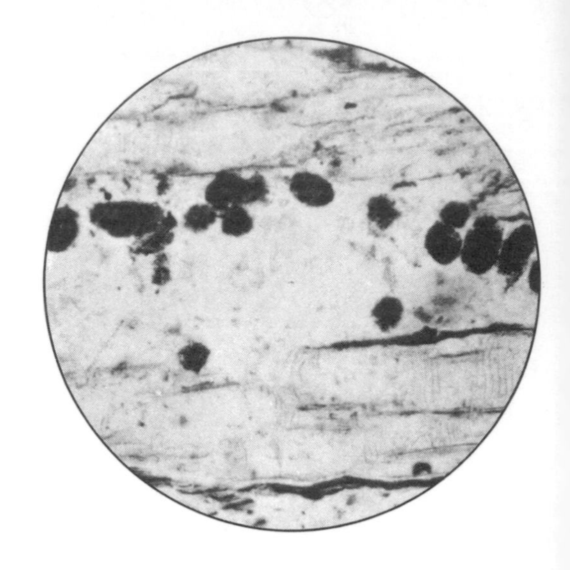

Below: four typical soil horizons. From left: chernozem, highly fertile with thick, dark zone of humus near surface; tropical forest soil, fertility reduced by heavy rains leaching organic matter from upper layers; podzol, infertile, its thin organic layer near surface strongly leached; desert, infertile, no humus, fine surface particles removed by wind.

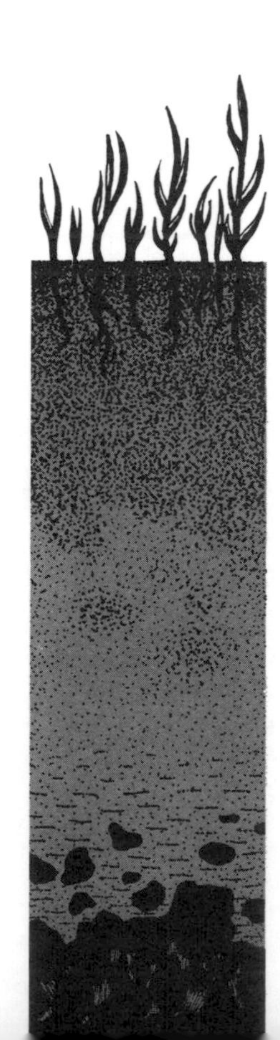

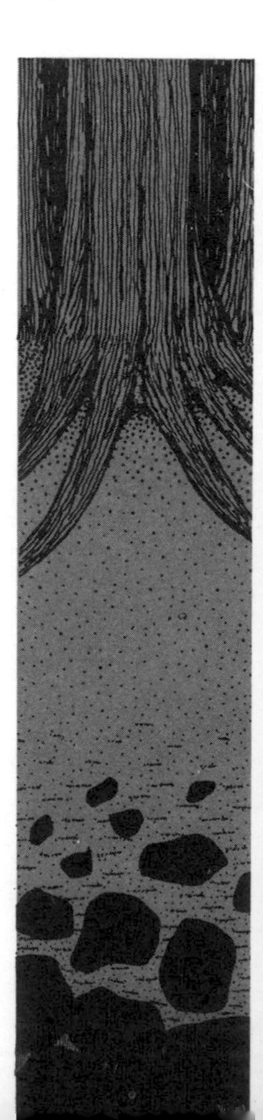

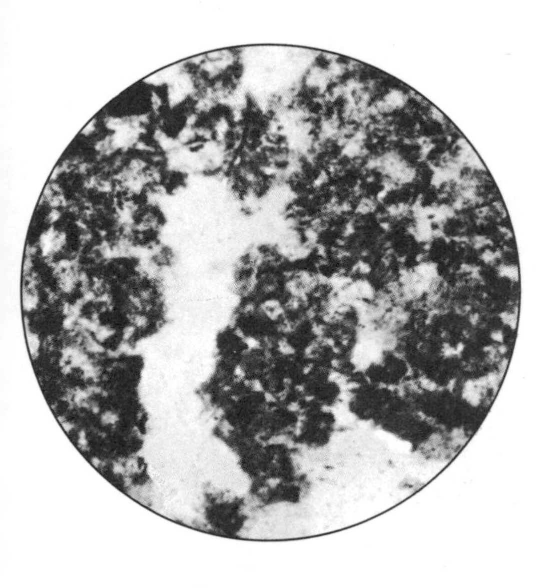

Photomicrographs (above) show two stages in the conversion of organic matter into humus in a brown forest soil. Left: dark, round blobs in the litter horizon, near the surface, are animal excreta. Right: in this lower horizon, excreta and other organic matter have decomposed; humus and mineral salts are intimately mixed. (Both photos ×100.)

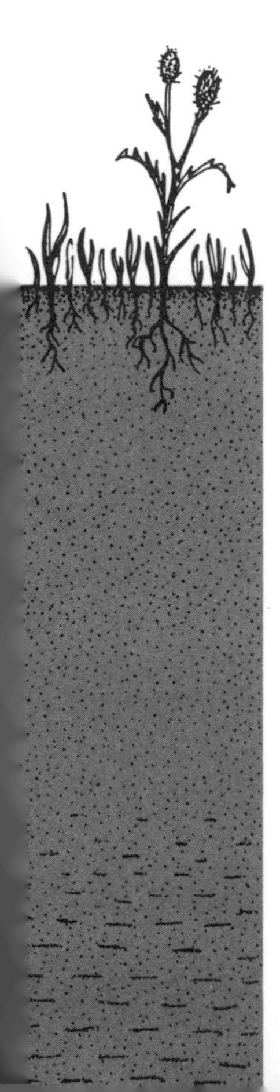

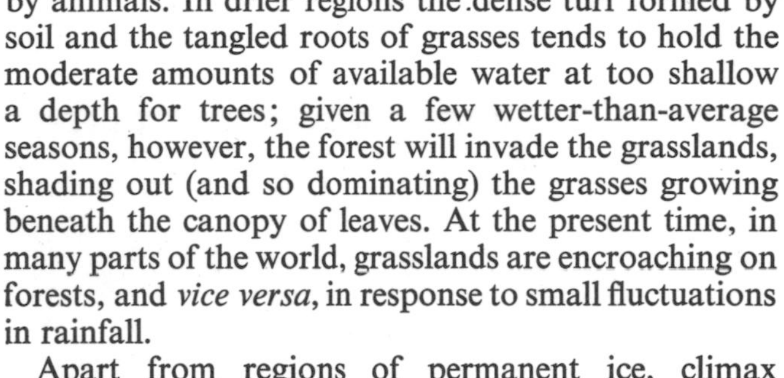

by animals. In drier regions the dense turf formed by soil and the tangled roots of grasses tends to hold the moderate amounts of available water at too shallow a depth for trees; given a few wetter-than-average seasons, however, the forest will invade the grasslands, shading out (and so dominating) the grasses growing beneath the canopy of leaves. At the present time, in many parts of the world, grasslands are encroaching on forests, and *vice versa*, in response to small fluctuations in rainfall.

Apart from regions of permanent ice, climax vegetation can be seen all over the world—in the meagre plant communities of the tundra and semi-desert as well as in the lush, dense forest of the tropics.

Just as each zone of vegetation is delimited by its climate, so it can also be distinguished by its soil. In the tropical forests soils often consist of red *laterites,* which are amongst the richest sources of iron and aluminium ores. Since heavy rains leach the humus out of their upper layers, laterites are usually poor soils for shallow-rooting plants such as grains and other food crops. The tropical grasslands receive less rain than the forest regions. Their typically reddish soils are thus better able to hold the humus at the surface, and their consequent fertility has been intensively exploited by the rice growers of south-eastern Asia.

Between the tropics and temperate regions lie most of the world's great deserts. Only a small proportion of each of these can be regarded as true desert in which nothing can grow. The greater parts of these regions are semi-desert and scrublands consisting of greyish, sandy soils. Although they lack rainfall, many of these soils could be made productive with the help of irrigation.

In certain temperate regions in the middle latitudes there are areas of black or dark-brown soils called *chernozems.* Their dark colour is due to their high concentration of humus, and they are amongst the most fertile soils in the world. Chernozems are found in areas of grassland where the grasses commonly grow to a height of two or three feet and form a thick turf of soil and matted roots as they die and accumulate. The fertility of these soils also owes something to the fact that the grasses tend to take up (and, later, deposit in the humus) a greater quantity of nutrients than do forest trees. Chernozems occur most widely in the North American prairies and in the Eurasian steppes. Almost equally fertile *prairie soils* occur in the so-called "mixed-prairie" regions—that is, in grasslands consisting of a mixture of tall and shorter

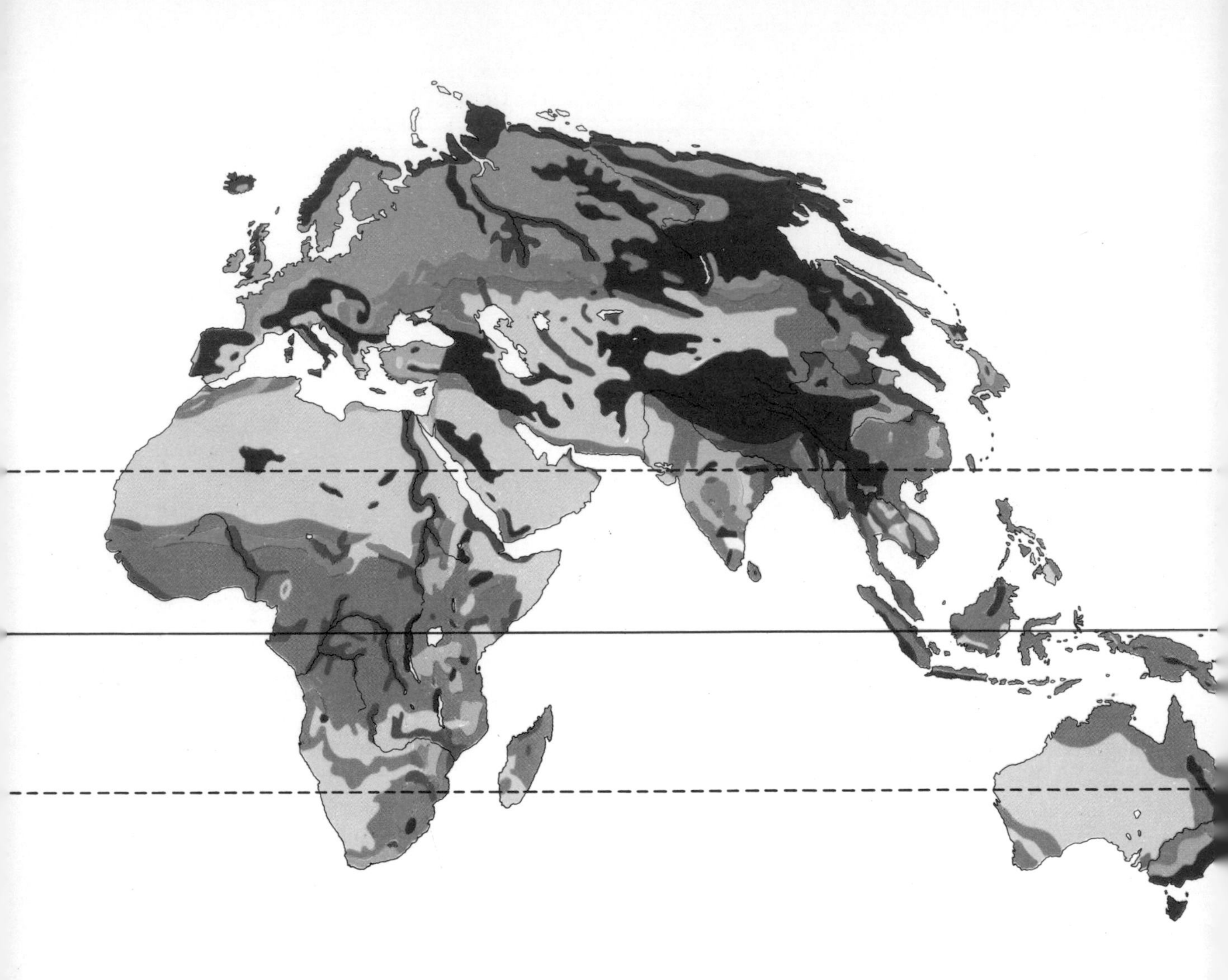

grasses. In North America, these soils occur on the eastern margin of the chernozems. Although containing a lower concentration of humus, prairie soils are at least as valuable agriculturally as chernozems because the climate is more humid and because the fertility of the soils extends to a greater depth. As a group, chernozems and prairie soils constitute the great wheat- and corn-growing belts of the Northern Hemisphere.

As we have seen, much of the land used for agriculture was originally covered by deciduous forests of oak, beech, and other trees. Since deciduous trees

World map indicates the main soil types. Soil forms when rocks at the surface are broken down into small particles by weathering. The inorganic soil fraction varies from place to place according to the nature of the parent rock. Fertility of soils depends ultimately on climate, which determines both the types of plant that can thrive in a given region, and the capacity of soils to produce humus and hold it near the surface.

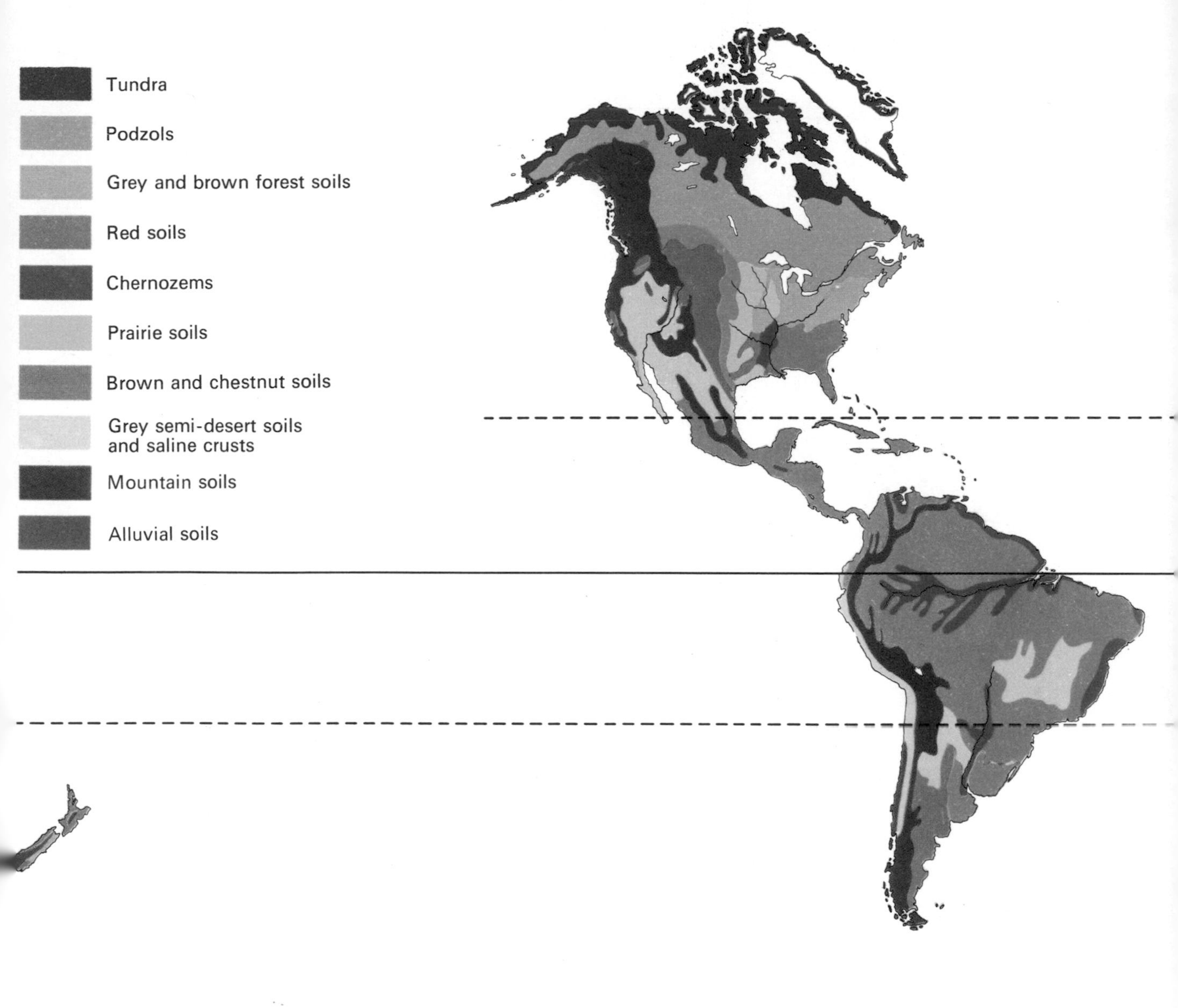

shed a heavy load of broad leaves every autumn, the *brown forest soils* are rich in humus and in mineral nutrients that their deep roots have absorbed from the lower soil horizons and from the parent rock beneath. In the temperate climates in which these forests flourish there is constant evaporation of moisture from the surface of the soil; this leads, in turn, to an upward migration of water from below by a process called *capillary action*, so that both humus and mineral nutrients are retained at the surface—a vital factor in the suitability of a soil for many shallow-rooting food crops important to man.

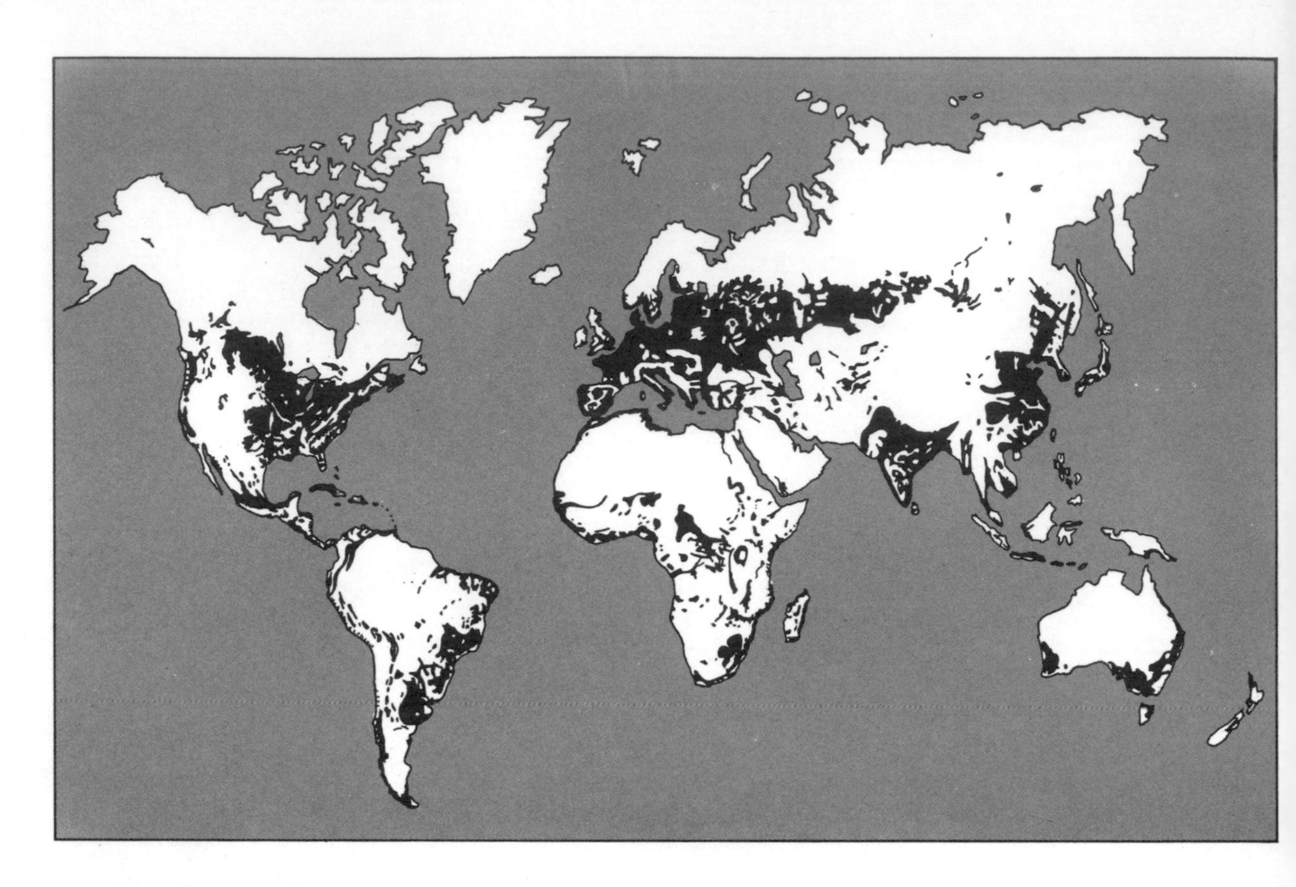

Land utilization map (above) indicates in broad outline the areas of the world at present cultivated for crops.

Agricultural land under permanent crops

Permanent meadows and pastures

Forest

Others — mountains, desert, tundra, etc.

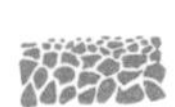

Photo-diagram (right) gives a more detailed picture of the extent of cultivated and uncultivated land (see key above). The width of the photographs indicates the relative acreages on a world scale.

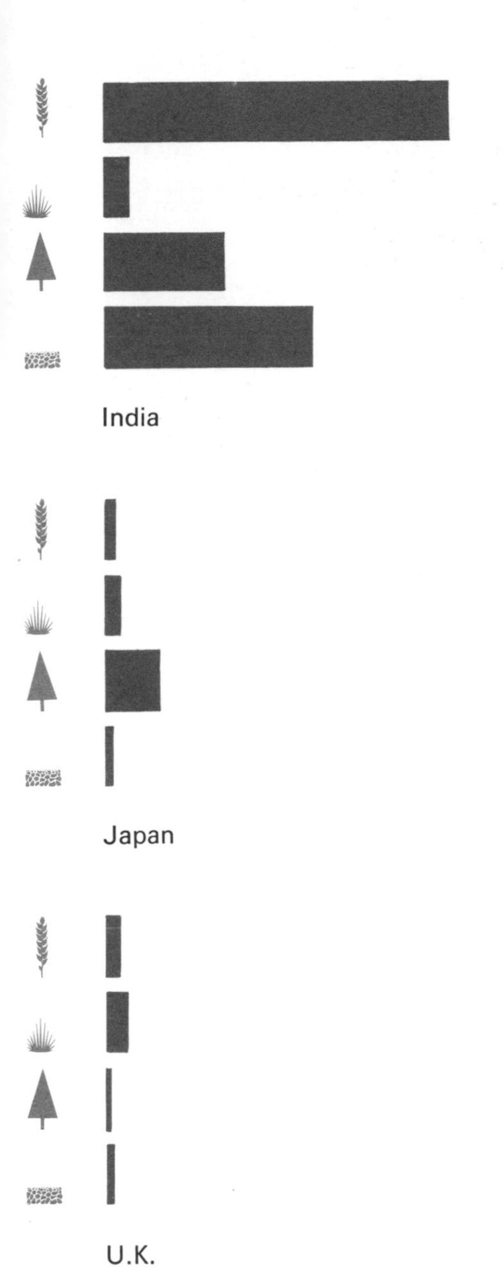

To the north of the deciduous forests of the Northern Hemisphere lie the great belts of coniferous forests, dominated by pine and other species. These trees are evergreens and in any case carry a smaller weight of leaf than the deciduous, so that they deposit much less humus. Moreover, there is little evaporation of moisture from the soil surface in these cooler latitudes, so that the net movement of water is almost invariably downward and the humus and nutrients carried by the water tend to become dispersed among the lower soil layers. Consequently *podzols*, as these soils are called, often consist of almost pure sand at the surface and are useless for agriculture unless constantly enriched with fertilizers.

Finally, to the north of the coniferous forests, lies the tundra with even poorer soils. These soils are often permanently frozen a few inches below the surface and support little plant life other than lichens, mosses, cotton-grasses, and other low-growing plants.

The soils we have described above are simply the principal basic types that occur from tropics to tundra. The dividing line between each is often blurred, especially in regions where one type of vegetation is encroaching on another in response to climatic variations. The suitability of a given soil for farming depends not only on the climate and the nature of the parent rock from which it was derived, but also on the existing cover of vegetation and, particularly, on the depth at which the soil concentrates its organic fraction. Unless a crop is grown on soil in which the humus and mineral nutrients occur at the correct depth, that crop will not thrive. In addition, of course, man radically alters the nature of the soil as soon as he begins to farm it. Once a soil has been cleared of its cover of, say, deciduous trees or grasses, it is denied its regular natural increment of humus, which must be replaced by artificial means—such as fertilizers or manure—if fertility is to be maintained.

Diagrams (left) indicate differences in land use in four countries. The scale of acreages is the same for all countries. Note the differences, from one country to another, between proportions of the available land devoted to grains, grassland, and forest; also the low proportion still uncultivated in the U.K., which is densely peopled and intensively farmed.

Lime (calcium carbonate) is spread on soils, as above, to reduce acidity. The most acid soils are peats and sands; least acid are chalk and limestone (in fact, pure chalk and limestone rocks are often used as liming materials). Acidity is measured on the *p*H scale, from 0 (highly acidic) to 14 (highly alkaline). Potatoes and oats are acid-tolerant crops, but even they will not thrive in soils with a *p*H of under 5. Clovers and sugar beet, with low acid-tolerance, need a *p*H of at least 6.5.

Science aids the Farmer

One of the ways in which civilization expresses itself is in the desire and capacity of individual people to specialize in one or more of an almost infinite variety of activities, either for pleasure or to earn a living, or both. Whereas in primitive tribes every able-bodied person is concerned in one way or another with hunting, gathering, or growing food, the proportion of an industrialized country's labour force working in agriculture is quite small and is growing smaller. Similarly, the amount of land available for food production for each member of the population is shrinking, and must continue to shrink as cities grow and populations increase. For instance, it is estimated that by the year 2000 the population of Britain will be 50 per cent greater than now but the area of land available for farming will have shrunk by a quarter.

This means that farming techniques must improve and that each acre of agricultural land must yield more and better crops. But it also means much more. For there is little point in sowing good seed if the maturing plant is at the mercy of pests and diseases; and there is equally little point in conquering such pests and diseases if the harvested food decays or is

Nitrogen is an essential constituent of all living tissues, and sufficiency or shortage of nitrates in the soil can make all the difference between high and low yields of every type of crop. The hay meadow (below) shows the effect of uneven applications of a nitrogenous fertilizer. (See also pages 85 and 86.)

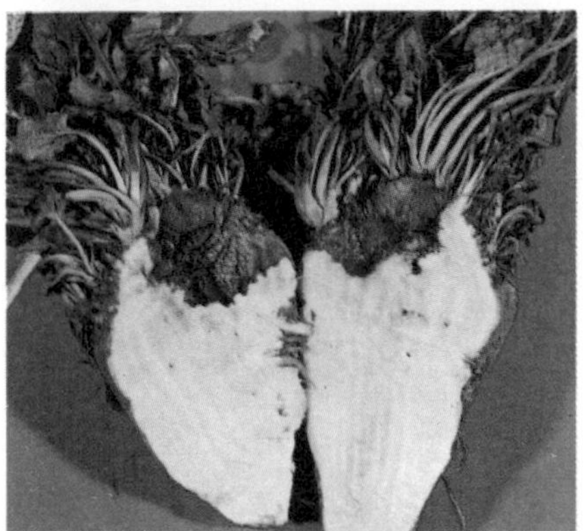

Though a plant needs some minerals in only very small amounts, their absence from the soil has drastic consequences. The cabbage (top) has a deficiency of potassium, resulting in withered leaf-edges; eventually, the entire plant will die back. In the sugar beet, heart rot (centre) is due to boron deficiency, while lack of manganese (bottom) causes its leaves to turn yellow and wither.

otherwise damaged during storage or on its way to the market-places of the world.

Throughout most of the history of agriculture, the farmer had only two methods of restoring fertility to the soil, unless he was fortunate enough to cultivate the seasonally renewed soil of a flood plain. He could dress his fields with animal and human dung, and he could apply the potash (potassium carbonate) and other nutrients that are present in the ash of burnt trees. It was only with the development of chemical fertilizers, which began in the early 19th century, that the farmer could be confident of maintaining soil fertility year in, year out.

One of the earliest discoveries was that heavy, acidic soils are improved by the addition of lime, which neutralizes the acid and adds the essential element calcium to the soil. It is widely used where clover, peas, beans, and other nitrogen-fixing plants are grown, since these cannot thrive in acidic soils. As we saw in Chapter 1, nitrogen is essential to all plants. At one time saltpetre (sodium nitrate), a mineral found in Chile, was the main source of nitrogenous fertilizers. In 1909, however, two German chemists discovered how to make ammonia (NH_3) by combining atmospheric nitrogen with hydrogen. Compounds such as ammonium nitrate and ammonium sulphate derived from the German process are now the basis of commercial nitrogenous fertilizers used all over the world. Potassium, which increases a plant's resistance to disease and drought, is available in the form of potash deposits that occur mainly in Germany and New Mexico, and need little processing. Phosphorus, which is important in the development of plant roots, is obtained by open-cast mining of phosphate rock (mainly calcium phosphate), a mineral found in abundance in Florida, in Morocco, on Nauru and Ocean Island (both in the Pacific Ocean), and in other parts of the world. Phosphate rock is treated with sulphuric acid to produce the fertilizer known as superphosphate—a form in which the phosphorus can readily be taken up by plant roots.

Nitrates, potash, and superphosphate are added to the soil in massive quantities. But the soil also needs much smaller doses of other nutrients. These *trace elements*, as they are called, include iron, manganese, copper, zinc, and boron, and all of them contribute to the growth of healthy plants. Soil dressings containing small quantities of one or more of these elements are now used by most farmers.

We tend to regard soil as a somewhat inert mass, but most healthy soils teem with an immense variety

of organisms, some harmful but most helpful to plants. Humus, as we have seen, is derived from plant and animal detritus that comes to rest on the surface of the soil, and from dead roots. None of this decaying matter can be transformed into humus until it is broken down by bacteria. These organisms inhabit the soil in almost unimaginably large numbers—perhaps as many as five or six million to the cubic inch—and no form of plant life could thrive without them. Of the larger inhabitants, earthworms are probably the most important. In eating, digesting, and excreting soil particles they not only improve the soil's fertility but also aerate it mechanically. Most important, they bring the results of their labours to the surface, where it is of greatest benefit to the growing plant.

In order to get the best out of his limited number of acres—whether he is raising corn, roots, or grass for pasture—the farmer prepares the soil in various ways. First he ploughs it—that is, he turns over the top few inches of the soil in order to aerate it and to bury (and so suffocate) any weeds that have invaded the seed bed. Next, he uses either a spike or disc harrow to smooth the surface and break up the larger lumps of soil—a task in which frost also lends a hand. Seed planting is nowadays done with drilling machines that deposit the seeds at the correct depth and, often, lay down fertilizers at the same time. Then the farmer may use a harrow to cover up the seeds and, finally firms the soil with a roller.

The plough has been a familiar and seemingly indispensable part of the farming scene for thousands of years. Yet, in some circumstances, ploughing may do more harm than good, for it not only turns the precious topsoil upside down but creates a hard platform at the base of the furrow. In short it seems to undo much of the good work of the worms—and, moreover, exposes them to hungry birds. For most purposes, at least on well-drained soils, soil aeration is done equally well by disc harrows, which both slice the earth and turn it, though to a much smaller depth.

Ploughing with oxen (above) and sowing seed by hand (below) in mediaeval times. Both methods were much slower than their modern counterparts, and so tended to limit the size of farms.

Right: typical modern farm machinery. The plough (1) overturns the topsoil, so allowing the air to get to it and burying weeds or stubble. The disc harrow (2) also turns the earth, but to a lesser depth, and breaks up lumps. The seed drill (3) plants seeds (and sometimes fertilizer also) at correct depth and interval. When shoots are an inch or two above the ground, the soil is firmed down with a roller (4).

1

2

3

4

Weeds can now be eliminated before seed planting by chemicals, such as paraquat, that kill green plants but are quickly neutralized by contact with the soil; they are thus harmless to worms and do not delay the planting of seed.

Chemical compounds like these are only a small part of a formidable array of substances available for protecting plants against a variety of natural enemies. These chemical substances fall into three main groups: *herbicides*, which kill weeds that compete with crop plants; *insecticides*, which kill insects that feed on plants; and *fungicides*, which kill fungi that cause many plant diseases.

These chemicals pose a dilemma because their use is bound to affect the exceedingly delicate balance that exists between the various plant and animal communities within even quite small areas. The control or eradication of, say, a plant pest may be vital to a farmer in the short term; but what if the only suitable insecticide harms his crop, or is dangerous to other animals (including man), or leaves toxic residues in the soil? Many insecticides, for instance, kill not only the pest but other insects that prey on it. It has often happened that a proportion of the pests survive the treatment and are then able to breed more rapidly than ever because most of their natural enemies in the neighbourhood have been destroyed.

The need for these chemicals is undoubted—provided they are highly selective and interfere as little as possible with the ecology of the area in which they are used. The immense difficulty of developing selective compounds is exacerbated by the capacity of many insects to develop resistance to chemicals within a remarkably short time. Meanwhile, other, non-chemical methods are also being exploited. One of the most interesting is *biological control*, in which a pest is fought by the deliberate introduction into the area of other insects that prey upon it. The classic example of this method is the use of Australian ladybirds that were imported to California during the late 19th century to control scale insects that had affected vast areas of citrus orchards. The ladybirds virtually eradicated the pest in two years. There are risks in biological control—the most obvious being that the "controller" will itself become a pest—and there is a limited number of species that can be used in this way. But it remains a promising field of research.

The protection of plant foods that are stored after harvesting is one of the most difficult problems in agriculture. The U.N. Food and Agricultural Organisation has estimated that about one twentieth of the

Above: the oats in the section at left were sprayed with selective weed-killer a month after planting; the section at right was not treated and is infested with charlock. Weeds are undesirable because they use water and nutrients needed by the crop and compete with it for light and air; they also add to the cost and difficulty of harvesting.

Below: five important enemies of crops. (1) Boll weevil, which attacks cotton; (2) Colorado beetle, a potato pest; (3) green peach aphid, which spreads diseases in potatoes and sugar beet; (4) red spider, which attacks fruit crops; (5) locust, which eats all green vegetation in its path.

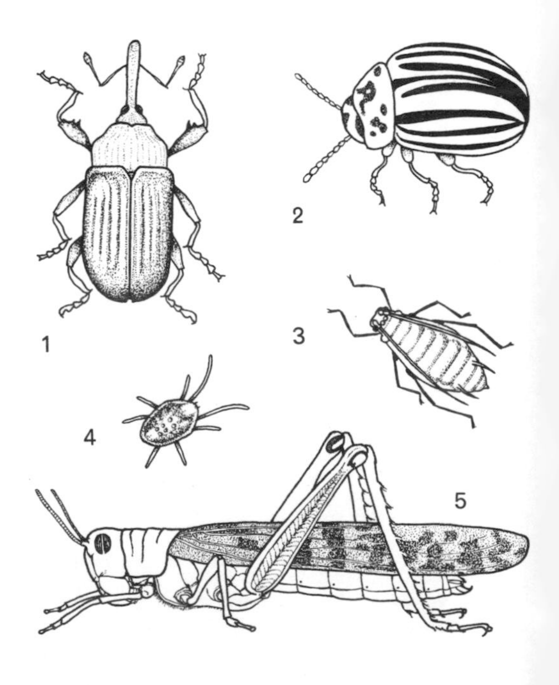

world's cereals are destroyed in storage every year; in some regions, the losses may rise as high as 50 per cent. The most serious pests are various species of beetles, which attack wheat, rice, maize, coffee, cocoa, nuts, herbs, dried fruits, and many other plants; the larvae of several species of moth also attack stored grains and fruit. Chemical sprays cannot be used to protect food because, even if they are not poisonous, they will impair its flavour. The best methods of protection are either to design granaries, silos, and warehouses so as totally to exclude insects, or to refrigerate the stored products, either in bulk or in cartons, for the retail market.

Our existing rates of food production and the sharply rising world population present us with a simple but frightening exercise in arithmetic. We often read about famines caused by crop failures in some part of the world; but we tend to forget that, even when freed from any immediate threat of starvation, about half the world's peoples are undernourished. According to the F.A.O., world food production will have to rise by 60 per cent during the next 20 years if we are to maintain even the present inadequate level of diet. And although new sources of food—such as the economic development of algae and other micro-organisms—and a much more scientific exploitation of marine life are likely to ease the problem in the more or less distant future, the immediate burden of feeding the world rests squarely on the shoulders of the farmer.

Above: crops are often attacked after harvesting as well as in the field. The apple (upper picture) was infected by brown rot, a fungus disease, after it had been packed. The wheat grains (lower picture) are being eaten by granary beetles after 16 months in storage. Every year, about five per cent of the world's stored grain is destroyed by these and other pests.

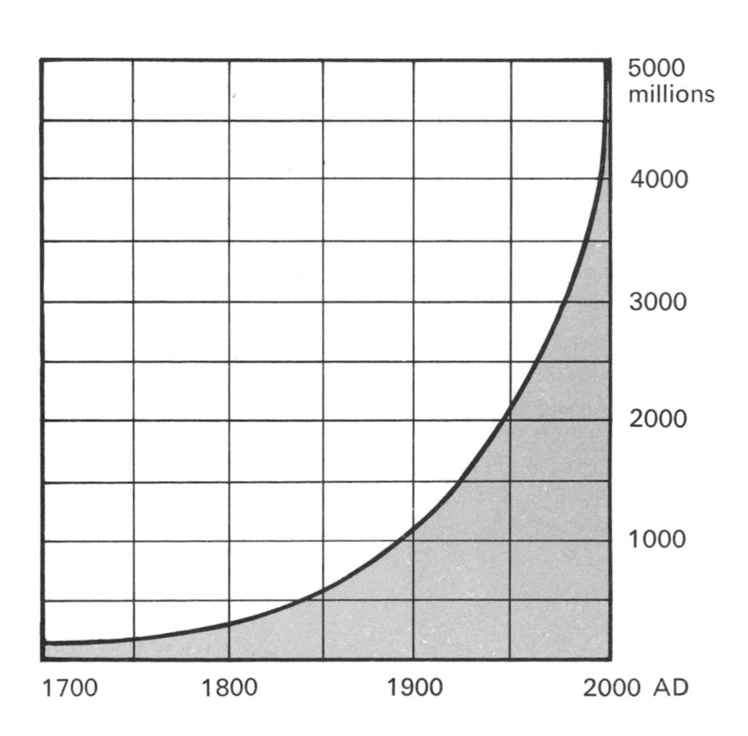

The graph (right) shows that the world population doubled between 1650 and 1850, and doubled again between 1850 and 1950. At present rate of increase it will again have doubled—to more than 5000 million—by the year 2000. Yet, even today, about half the world's people are undernourished.

3 Grains

All the great civilizations of the past and present have been founded upon agriculture based on grain crops, or cereals, as we call them. The ancient cultures of the Near and Middle East, Greece and Rome in the classical period, and subsequent civilizations of Europe and America were (and are) dependent on wheat, and to a lesser extent on maize, barley, rye, and oats. Maize provided the staple diet of the peoples of the old Indian civilizations—Inca, Mayan, and Aztec—of the New World. And from the earliest times China, India, and the other civilizations of the Orient have been based on rice.

This dependence on cereals is due primarily to the high food value of their seeds. A cereal grain is rather like the egg of a chicken: it contains not only the embryo of a new plant but also a rich store of food that will sustain the embryo until its own food-producing apparatus (roots and leaves) has had time to develop. All the grains that we value as food contain carbohydrates, proteins, minerals, vitamins, and a little fat. As food for humans they offer a more balanced diet than any other plants. Indeed, millions of people in Asia live—though not well—on rice and little else, while maize plays an almost equal part in the diet of many South American Indians.

The grain plants are all species of grasses that once grew wild. Most of them, like other grasses, are *self-fertile*: their inflorescences, or flower clusters, contain both *anthers* (male organs) that produce pollen, and *stigmas* (female organs) that receive the pollen and produce the seeds, or fruit. Since these plants need no outside help in pollenization, they do not have the bright-coloured or highly scented flowers that other plants develop in order to attract birds or insects.

The cereals all differ in one important respect from the typical grasses one sees in meadow or garden. For the pasture grasses are *perennials*: they scatter seeds every autumn, but their lower stems and roots continue to grow and will develop new leaves, flowers, and seeds the following season. The cereals, on the other hand, are *annuals*, the plants dying when their

Vase depicting three maize gods is a product of the Mochica peoples (first-ninth centuries A.D.), who used irrigation techniques to grow maize in the large valleys of northern Peru.

Wheat is the most widely cultivated food plant in the world and provides the daily bread of countless millions of peoples in Europe, the Americas, and much of Asia.

life cycle has reached its climax with the development of seeds. It is odd, at first sight, that man should have chosen as food crops those grasses that must be sown afresh each season. The fact is, only the annuals develop seeds that are sufficiently large and nutritious to make harvesting worthwhile. In the wild, the survival of an annual, in contrast to a perennial, depends solely upon the embryo produced each season; its massive seed, crammed with food, is the plant's insurance against extinction.

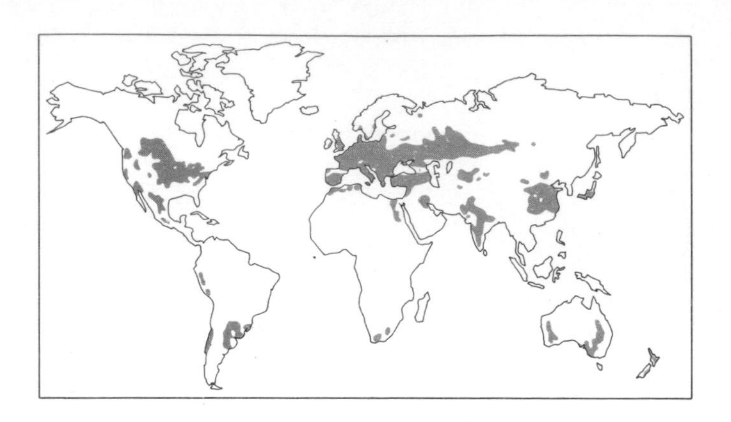

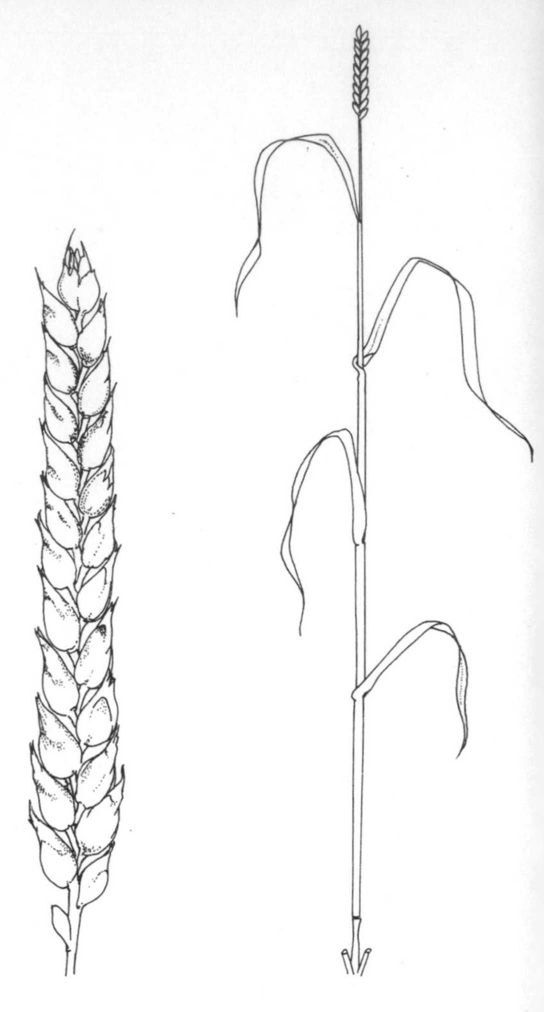

Wheat

Every year some 400 million acres of soil are devoted to the production of wheat. Wider in geographical distribution than any other cultivated food plant, wheat is the most important food crop in the temperate zone, and is also grown in almost every country from the equator to the Arctic Circle.

It is also, perhaps, the oldest of man's domesticated cereals. Some years ago, archaeologists exploring the site of the 6700-year-old village of Jarmo, in eastern Iraq, unearthed a number of charred grains of wheat. When samples of modern grains were also charred to enable comparisons to be made, it was found that the ancient grains were almost identical with two types of wheat growing today—one a wild species, the other a domesticated species called einkorn. Jarmo is one of the oldest villages yet discovered in the Tigris-Euphrates basin and may have been one of the birthplaces of agriculture; so there has been little change in these wheats since man first domesticated them.

All the known wheats—wild and domesticated—are classified as separate species of a single genus, *Triticum*. Botanical opinion differs as to the exact number of species, but many authorities agree with the Russian geneticist Nikolai Vavilov, who recognized 14 species. Wheats can be divided into three groups, each distinguished by the particular number of chromosomes in their cells. Chromosomes, which occur in the cell nuclei, are responsible for the transfer of genetic material from one generation to the next. Each group of wheat species has 7, 14, or 21 chromosomes in their reproductive cells, and these numbers greatly influence the appearance, structure, resistance to disease, and other properties of the group. The 14-chromosome wheats are all derived from 7-chromosome species (einkorn or wild einkorn) crossed naturally with related wild grasses by a process that involves chromosome doubling. The 21-chromosome

Map (above left), shows the principal wheat-growing areas of the world. In addition, wheat is cultivated in rather smaller quantities in many other parts of Europe, Asia, and South America. Yeoman (above), an autumn wheat, is one of hundreds of bread-wheat varieties developed by hybridization.

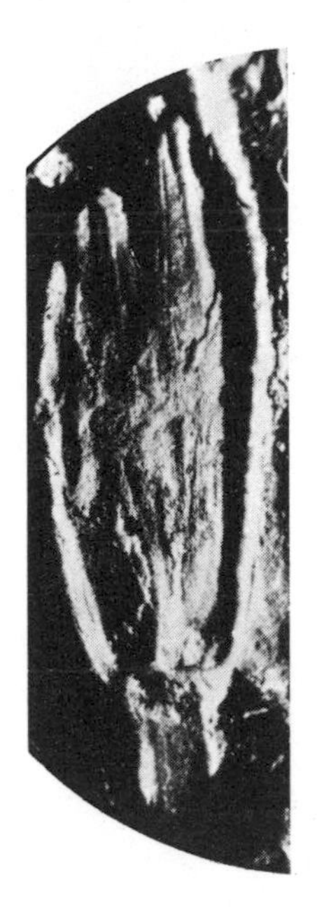

Above: cast of a spikelet of wheat found at Jarmo, Iraq (left), and a spikelet of modern emmer (right). A comparison suggests that this type of wheat has changed little in the 7000 years that separate the two samples.

species are hybrids of 14-chromosome wheats and 7-chromosome wild grasses, again with the help of chromosome doubling.

Seven-chromosome einkorn (either wild or domesticated) is thus responsible, directly or indirectly, for all the other species of wheat, including those cultivated at the present time. For example, common wheat—the most widely grown species—is thought by some to have derived by chance from crossing of 14-chromosome Persian wheat with a 7-chromosome wild grass (*Aegilops squarrosa*) that commonly infests wheatfields from central Europe to western Asia. In turn, Persian wheat developed as a hybrid of einkorn and a 7-chromosome wild grass. It is certain, then, that wild or domesticated einkorn antedated the other species. But hybridization and chromosome doubling *can* take place very rapidly, so there is no particular reason why even the 21-chromosome species may not be almost as old as einkorn. Seeds of 21-chromosome shot wheat, for example, have been found at the site of the ancient city of Mohenjo-Daro, in West Pakistan, which is at least 4500 years old. It is also certain that all the original species of cultivated wheat arose spontaneously, man's only contribution being unwittingly to encourage hybridization by growing wheat in many different parts of the world, and so increasing the range of possible wild-grass hybrid parents.

Today an enormous range of wheats is available to the farmer. They are not new species but are *varieties* that have been developed, by planned hybridization, from common and other wheats. The aim of hybridization is to concentrate as many desirable properties as possible in a single variety. One variety, for instance, may have good milling properties but poor resistance to diseases such as wheat rust; another may resist rust but provide a low yield of grain per acre; a third may give a high yield but bake poorly because of low concentrations of gluten in the grain; a fourth may give good baking flour but succumb to dry weather conditions in the field; and so on. New varieties are constantly being developed by plant breeders. To the basic properties mentioned above are added others that vary according to the climate in which the wheat is to be grown and whether it is to be planted in the autumn or spring. The improvement in varieties is largely responsible for the enormous increase in world production of wheat during the last 30 years. During this period, while the area under wheat in the main producer countries has increased by only a fifth, the yield per acre has increased by about a third, and total production by more than half.

Below: most authorities agree there are 13 wheat species in addition to common, or bread, wheat. The six shown here are (left to right): 7-chromosome einkorn; 14-chromosome emmer, macaroni wheat, and Polish wheat; 21-chromosome club wheat and spelt. All are cultivated today.

Wheat grows best in regions that receive less than 30 inches of rain a year. It needs warm weather for the ripening period, but it also requires a short cold season during early stages of its growth to discourage the development of diseases such as rust, smut, and mildew. The ideal soil for wheat is a fine, crumbly tilth that is nevertheless firm and free from clods. A good mellow soil of this kind is achieved only by ploughing well in advance of sowing and allowing the soil to weather slowly. Farmers often plant wheat on soil in which potatoes or roots have been grown the previous season, since both these crops leave the soil in a weed-free, fertile, and well cultivated condition; indeed, it may be *too* fertile for less demanding cereals such as barley or oats, which might grow too rapidly and collapse in high winds. The *rotation* of crops is important for two reasons: first, it helps to prevent the soil from becoming exhausted; second, it discourages diseases (which usually attack only one species or group of related species) from establishing themselves in a particular area of cultivated land.

Wheat, like other crops grown from seed, is nowadays planted by a drill instead of broadcast by hand. Apart from being quicker and less wasteful, the drill is able to place the seed at the correct interval (about eight inches apart) and at the depth required (about one and a half inches); and since all the seeds are at exactly the same depth, the shoots will all emerge above ground together and, later, their heads of corn will ripen at the same time. The amount of seed planted varies, the cooler, wetter areas like northern England and Scotland needing a greater quantity of seed than the warmer, drier areas like, for instance, the southern prairies of the United States. As a rule the rate varies between 75 and 150 pounds of seed to the acre.

Almost every grain sprouts, but less than half (and often only one third) of them will grow to maturity; the rest are eaten by birds, mice, and insects, or succumb to fungal diseases or spells of extreme weather. Early growth is rapid because, as we have seen, the seed contains an abundant store of food. About six weeks after sowing, when the small plant is becoming dependent on photosynthesis and absorption through its roots, the farmer may enrich the soil with a mixture of nitrate, phosphate, and potash. He will also apply a selective weed killer to destroy annual weeds that germinated at the same time as the wheat.

At first, the young wheat plant resembles other grasses. Shoots soon develop at nodes on the main

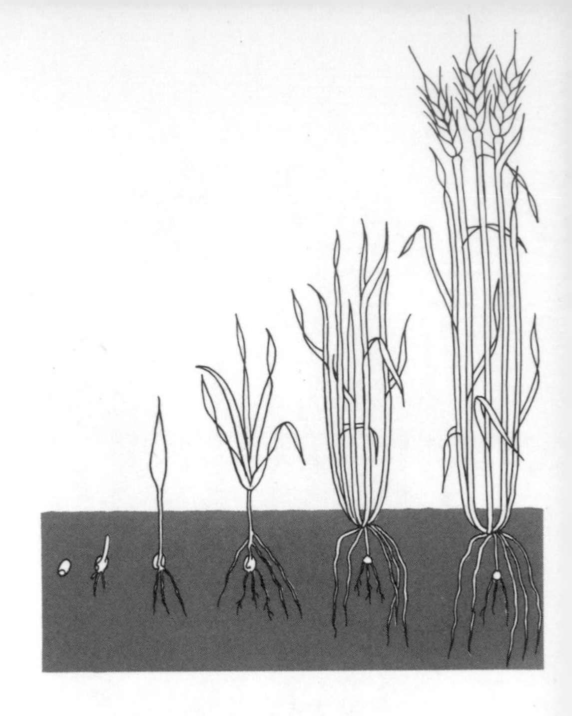

Diagram (above) and photo (right) show development of tillers in the young wheat plant. The yield of grain in the wheat crop depends to a great extent on the proportion of tillers that have produced mature ears by harvest time.

stem that are immediately below the surface of the soil. These shoots, called *tillers*, give rise to several stems and, thus, several heads of grain from each plant. At the same time, roots develop from the subsurface nodes to supply the tillers with water and nutrients. At first, only the leafy blades of the main and tiller shoots are visible above ground. As summer advances, the shoots grow taller, eventually reaching a height of about three feet, and develop compact green flower clusters at their heads. Wheat, as we have seen, is self-fertile: each plant is usually fertilized by its own pollen, though sometimes pollen wind-blown from neighbouring flower clusters may contribute. Almost every seed develops in each head of grain. The seeds stand in straight rows around the central stem, each enclosed in a papery husk.

As they ripen during August and September, the plants gradually change in colour from green to golden brown, and soon after are ready for harvesting. The knowledge and experience of the farmer are vital at this stage. The ripe grains, as we know, are filled with foods that have, so to speak, been fed into them from the rest of the plant. If the crop is cut too early, the seeds will lack their full complement of food; if it is cut too late, the grain will yield a poor flour owing

Below: wheat, like many other grains, is nowadays cut by combine harvesters that remove the ears from the stem and thresh the grain out of the husks in one operation. Although much quicker than older methods, the combine should be used only if the crop is dry because damp grain will deteriorate in storage.

Above: breadmaking and brewing were closely linked crafts in Ancient Egypt. The model (about 1200 B.C.) shows pottery casks for storing beer.

Below: the mediaeval painting (left) shows bakers removing a loaf from a brick oven. In a modern bakery (right) production is highly automated and the bread untouched by hand at any stage.

to thickening of the *bran*, or outer covering of the kernel.

One of the most important differences between modern bread wheats and the primitive cereals and grasses from which they are derived concerns the grain heads. The heads of old and wild wheats have central stems that are brittle and will break easily when the plant is mature, while their seeds may remain firmly enclosed within their husks even when threshed. These properties help in the dispersal and survival of the seeds in the wild, but create difficulties in a cultivated plant. Modern bread wheats, by contrast, have strong stems that will stand up to the wind, so that their heads remain intact even when mature; moreover, threshing readily parts their seeds from the husks. Thus, the modern farmer can count on losing only a small proportion of his grain crop during the various stages between reaping and storage. In a normal year the harvest will yield, as saleable grain, between one and two tons an acre.

Today, almost all the wheat produced in the wealthier countries is used for making bread. But it is extremely unlikely that the earliest farmers used it in this way. Probably, they heated the grains, making it easier to detach the husks and chew the kernels within. The discoveries in the dwellings at Jarmo give weight to this assumption. A little later the farmers began to grind the parched grains into a coarse meal and soak it in water, producing a primitive form of porridge. If the gruel was left for a few days in a warm atmosphere

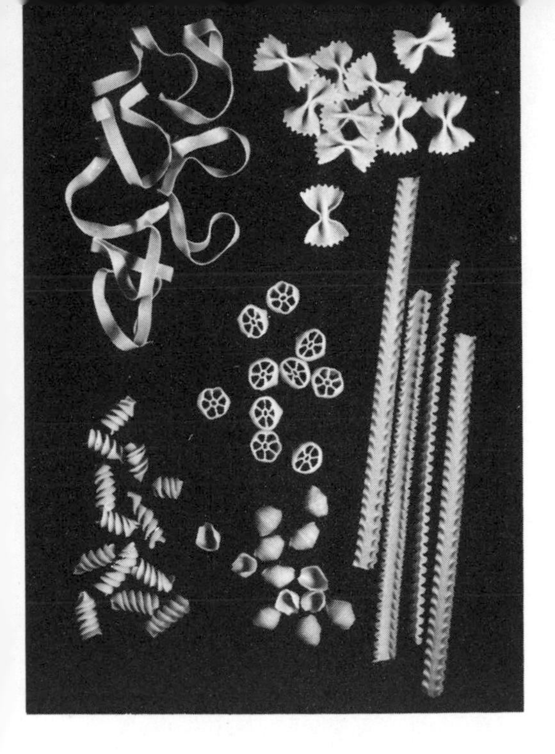

Above: a few of the immense variety of paste products made from durum wheat.

Below: several of the world's major wheat-growing countries regularly produce a surplus. The map shows the general flow of the international trade in wheat. One of the main problems in feeding the undernourished peoples of the world lies in persuading them to eat cereals to which they are unaccustomed.

it would be invaded by wild yeasts that would ferment the sugars in the wheat, producing an alcoholic beverage. In such a way, perhaps, the secret of making leavened bread was discovered. Certainly, bread-making and brewing, both dependent on yeasts, have close historical links, and we know that at least 5000 years ago the Egyptians were making a beer-like drink by fermenting half-baked bread.

Wheat destined for the modern bakery is ground between massive steel rollers that are set either to crush it fine or to remove only the bran. If the whole grain is ground fine it produces *wholemeal*, a brown, very nutritious flour with a strong flavour. More often, the bran is removed (for use in animal feeds) and the remaining kernels, rich in starch but lacking the extra protein contained in the bran, are ground to produce the familiar white flour.

Apart from common wheat, commercially the most important species is the 14-chromosome *Triticum durum*, which is used in the manufacture of spaghetti, macaroni, and other paste products. It is grown mainly in Spain, Algeria, southern U.S.S.R., and India.

Of the present world wheat production of about 250 million tons a year, the Soviet Union produces some 70 million tons, the United States 30 million, China 25 million, and Canada 20 million. India, France, Turkey, Argentina, and Italy all produce more than eight million tons, while Pakistan, West Germany, Spain, Yugoslavia, Poland, Romania, and the United Kingdom produce four million tons or more.

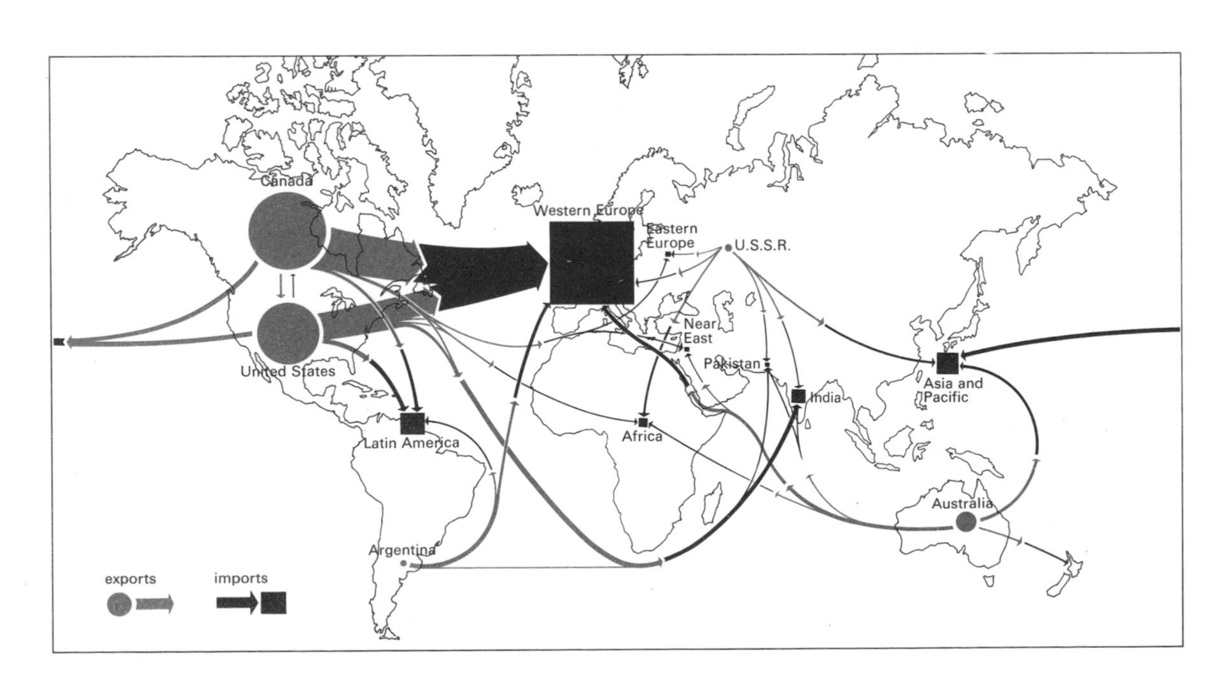

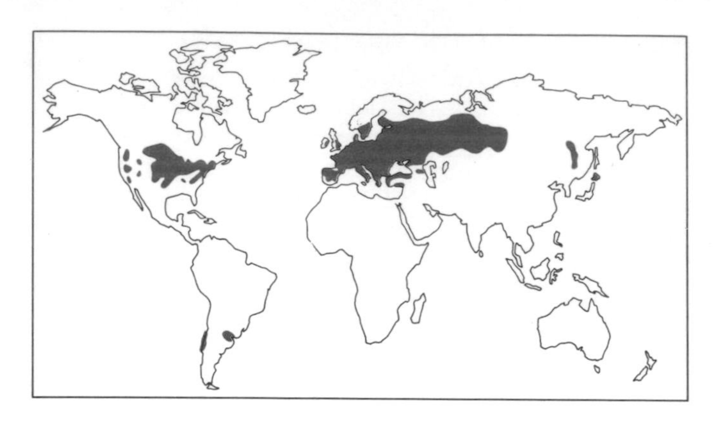

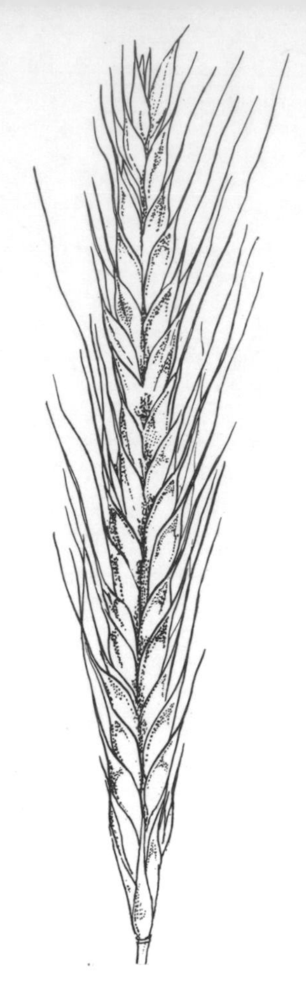

Map (above, left) shows the main rye-growing areas of the world. Self-sterile, the rye plant (above) will produce a head of seeds only if it receives pollen from one of its neighbours in the field.

Rye

In its domesticated form rye is a comparatively modern cereal, our earliest records of its cultivation dating from the Roman period. It appears to have arisen from a wild perennial still common in south-eastern Europe and in Asia Minor. Unlike wheat, oats, and barley, rye is *self-sterile* and must be cross-pollinated; in other words, an individual rye plant will not produce seeds unless it receives pollen from one of its neighbours.

Although, on a world scale, it is much less important than wheat, rye is a useful crop because it grows on poor, light soils that would be unsuitable for other cereals. Moreover, it is hardier than wheat, oats, and barley and will tolerate lower winter temperatures and higher soil acidity. Cultivation, sowing, and harvesting methods are similar to those for wheat. The rate of seeding is from 100 to about 170 pounds to the acre, and the yield is about a ton an acre.

Although rye is used primarily as a bread crop, the winter varieties provide a useful bonus for the farmer who keeps livestock. Sown in the early autumn, winter rye grows vigorously during the remainder of the year and by early February furnishes an excellent pasture for sheep and cattle. As long as the animals are not allowed to bite the crop too short, it recovers quickly and will later produce a normal yield of grain. The mature grain, richer in starch than wheat, is greyish brown in colour and yields a coarse dark flour with a distinctive nutty flavour. At one time rye was the principal raw material in the manufacture of vodka; today, however, it is distilled mainly from potatoes and maize, to which small quantities of still-green rye have been added.

World production of rye is about 35 million tons a year. About 93 per cent is grown in Europe, with the Soviet Union producing half the total, Poland one fifth, and West Germany one tenth.

Below: the typical rye loaf is darker and coarser textured than wheat bread.

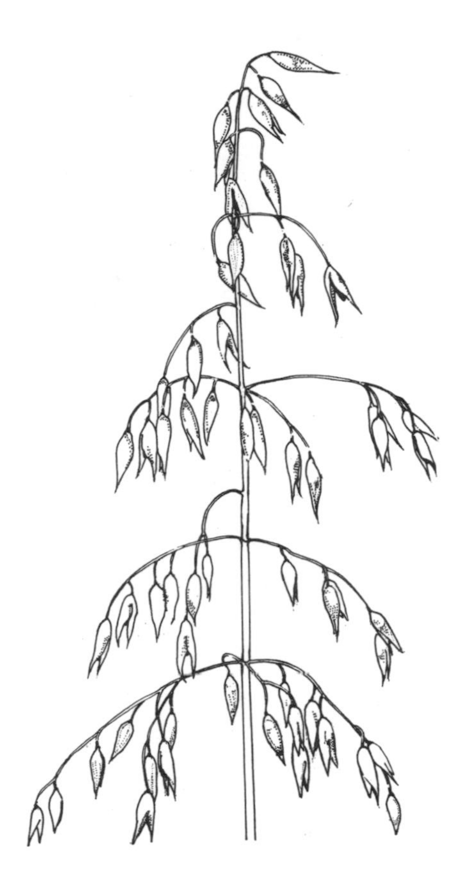

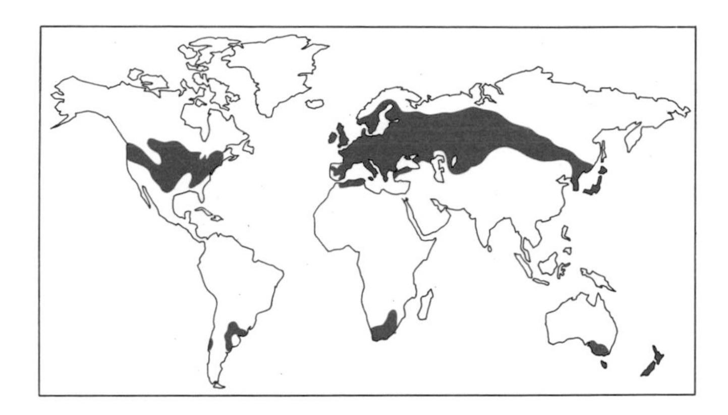

Map (above, right) shows the main oat-growing areas of the world. The oat ear (above) consists of seeds hanging from widely separated branches on the stem.

Below: the Tartarian oat (left) has a dense cluster of seeds growing from branches on one side of the stem. Smaller seeded, and with longer awns, the variety called Ceirch Llewd (right) is grown on the Welsh hills.

Oats

Less is known about the wild ancestors of the oat than of most other cereals. The Old Testament, a useful guide to pre-Christian farming practice, makes no mention of it; indeed, any information about it before the mediaeval period is hard to come by. It is likely, however, that it developed (like wheat and barley) from wild grasses growing in the Near East.

The oat is easily distinguishable from wheat, rye, and barley because it develops an ear in which the spikelets containing the seeds are not clustered compactly about the central stem but droop from branches that arise from widely separated nodes on the stem. The oat genus *Avena* consists of four main cultivated species. The most important commercially is *A. sativa*, the common oat, in which each spikelet produces two or more ripe kernels enclosed within a thick husk. There are many varieties of common oat; they are grouped according to the colour of their husks, which may be white, yellow, grey, or black. The rate of seeding is between one and two hundredweight per acre and the yield is from one to two tons an acre. The Tartarian oat (*A. orientalis*) is easily identifiable because its spikelets arise from branches that grow only on one side of the central stem. The naked oat (*A. nuda*) is confined mainly to the upland regions of China. Its spikelets bear either six or seven grains, and the kernels are much more easily freed from the husks than those of the common oat. The bristle-pointed oat (*A. strigosa*), cultivated in the Welsh hills, has a smaller grain than the common oat and develops a long *awn*, or beard. *A. fatua*, the wild oat, is an extremely troublesome corn-crop weed.

Europe (notably Poland, France, West Germany, the United Kingdom, and Sweden) and North America each produce about the same quantity of oats, their combined total accounting for about 98 per cent of the world production of some 50 million tons a year.

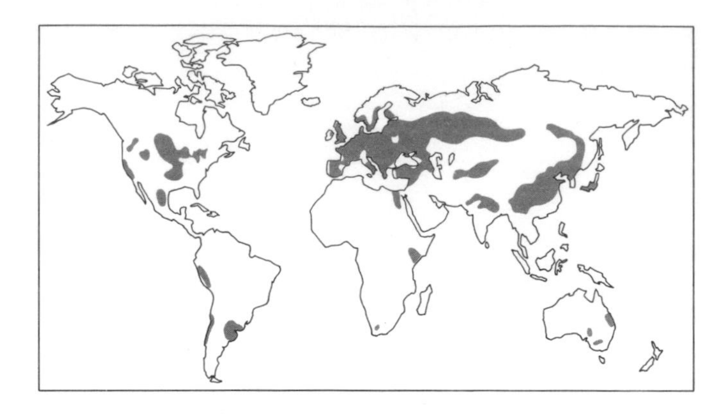

Drawing (above) is of two-rowed barley, with two rows of seed-bearing spikelets and, in the centre, two rows of sterile spikelets (the other two sterile rows being on the opposite side of the ear). The map (above, left) shows main areas of cultivation of both two-rowed and six-rowed barleys.

Barley

Barley was cultivated (with wheat) by the earliest farmers of the Middle East and may have been the first cereal to be domesticated by Europeans. Barley belongs to the *Hordeum* genus of grasses, and it is likely that it developed from *H. spontaneum*, a wild grass still found in the Caucasus and elsewhere in western Asia.

There are many species of barley, but only two types are economically important—six-rowed and two-rowed. These type-names refer to the ways in which the spikelets develop on the ear. In six-rowed barley, three fertile spikelets—one central, or median, and two lateral—form at nodes on opposite sides of the main stem. Two-rowed barley is so called because, although the spikelets develop in a similar manner to the six-rowed types, only the median spikelets are fertile and produce seeds. (A six-rowed species known as Bere barley is sometimes called four-rowed because the spikelets appear to be arranged asymmetrically in four rows.)

Although it is often grown as part of a rotation, after either roots or wheat, barley can flourish on lighter soils than wheat or oats because its roots are thin and do not strike so deep. This, coupled with the fact that it requires little more than two months between sowing and harvesting, has led to its cultivation not only in temperate regions but also at quite considerable altitudes in the tropics. Barley's main weakness—a low tolerance of acidity—can be overcome by adding lime to the soil, preferably to the preceding crop in the rotation. Cultivation and sowing are similar to those for wheat. The seeding rate is about 100 to 150 pounds to the acre, and the yield average from one and a quarter to two tons—about the same as wheat.

Although much barley is grown as a cattle food, its most important use is in brewing beer and distilling

Right: barley is laid out on the floor of a malting shed (top) of a malt-whisky distillery. Here the grains germinate and convert some of their stored starch into sugars under controlled conditions. In the mash tun (centre) warm water is added to the malt to complete the starch-conversion process. The sugar is then fermented with yeasts. In the pot stills (bottom) the alcohol is separated from the water by boiling, the alcohol vapour being led off through the pipes in the top of the stills.

whisky. For brewing, barley's quality depends upon a high concentration of carbohydrates and a low level of protein—preferably below 1.3 per cent of the weight of the grain. Since protein forms during the ripening of the grain, and since growth (including ripening) of barley is rapid, the timing of harvesting is critical to the production of high-quality beer.

The first stage in the brewing process is *malting*, in which the threshed kernels are steeped in water and then spread out on a tiled floor in a warm atmosphere and left for about ten days. During this time the kernels will begin to sprout. The point of this process is that, in order to sprout, the kernels must first convert much of their store of starch into sugars, which are essential to brewing because, unlike starch, they are soluble in water. When sprouting has reached a certain stage it is arrested by drying the malt in a kiln. Next, the malt is crushed and immersed in hot water. The resulting liquid, which consists of the malt sugars dissolved in water, is called *wort*. It is separated from the solid matter and then boiled for several hours. Now the cone-shaped fruits of hop, a perennial climbing plant, are added. Hops give beer its bitter taste and distinctive aroma and also act as a preservative and antibiotic.

After boiling and filtering, the wort is fermented with yeasts. Yeasts are enzymes that catalyse a reaction in which the malt sugars are converted into alcohol and carbon dioxide is released. The effervescent, or bubbly, quality of beer is due to its retention of some of the carbon dioxide; in the case of some lagers and other beers, effervescence is increased by "carbonation" (the addition of carbon dioxide) immediately before bottling.

Whisky (that is, the spirit made in Scotland; the Irish and American equivalents are usually spelt "whiskey") is made by distilling the wort. Although the art of distilling (as of malting and fermenting) is complex and is modified locally by tradition and various "tricks of the trade," it consists basically of boiling the wort in order to produce a concentrated liquor. Since alcohol boils at a lower temperature than water (78.5°C as against 100°C) it is collected as a vapour and then condensed. Outside Scotland, whiskeys are often made from a variety of cereals. Some Irish whiskeys, for instance, are made from a mixture comprising 90 per cent barley and 10 per cent oats, wheat, and rye. In the United States corn whiskey is made from maize alone, rye whiskey from rye alone, and bourbon from 65 per cent maize, 20 per cent rye, and 15 per cent barley.

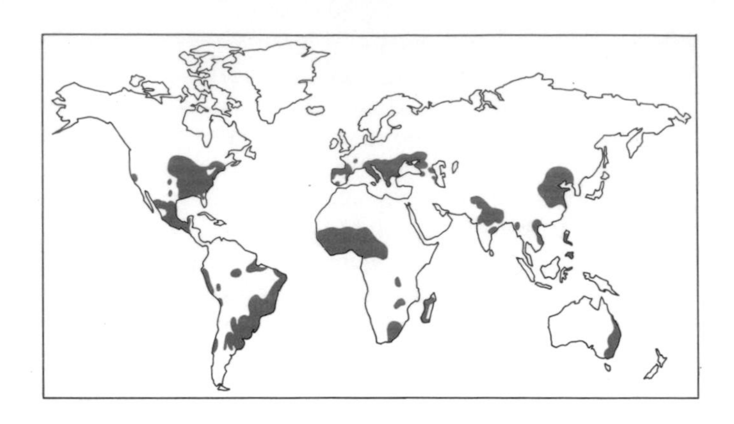

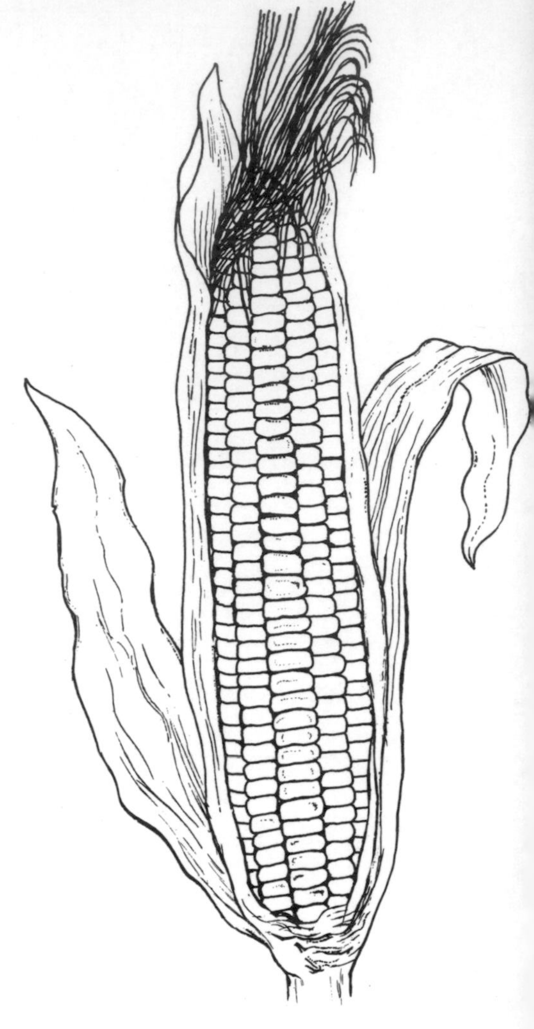

Maize

In all the cereals we have considered so far, the modern domesticated species are not substantially different in their botanical features from their wild ancestors. True, the size of the grains has been increased, and many of their economically important properties have been improved. But in appearance and structure they bear close and obvious relation to the wild grasses from which they are derived.

Maize, which is indigenous to America and nowhere else in the world, provides an astonishing and still somewhat mysterious contrast to the gradual and relatively modest evolutionary development of these other cereals. The earliest samples of maize ears yet discovered (in Bat Cave, Mexico) are about 4000 years old. Superficially, they bear as much resemblance to an ear of wheat as they do to the massive maize cob familiar to us today. The seeds of these ancient ears were each enclosed in husks and were, if anything, even smaller than those of wheat. Like einkorn, these primitive maize ears were well adapted for seed dispersal and survival in the wild. Most modern cultivated maize, on the other hand, has become so

Map (above, left) shows that maize, though indigenous only to the Americas, is now cultivated in many parts of the world. The seed head, or cob, of the maize (above) is considerably larger and has many more seeds than other grains.

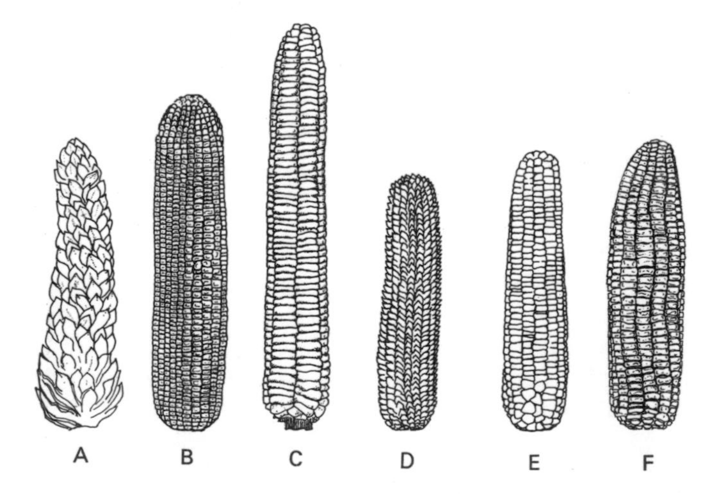

Left: the six principal types of maize. (A) pod corn, a probable ancestor of modern maize; (B) dent corn; (C) flint corn; (D) pop corn; (E) flour corn; (F) sweet corn.

specialized that it probably could not survive without the help of man. The glumes have almost disappeared and the whole cob—which may be as much as two feet long and bear many hundred tightly adherent seeds—is now protectively enveloped in leaf-like husks. If a ripe cob falls to the ground, the chances of the seeds surviving are slight because they would be competing with each other in too small a patch of soil to obtain sufficient moisture or nutrients.

From what species of maize, then, have our modern plants developed? The short answer is: we don't know for sure. It is, however, certain that they did not develop from some mysterious, undiscovered ancestor of comparable size and shape; rather they have developed (very quickly in evolutionary terms) from much smaller wild and domesticated relatives that are still common in the Americas. At one time it was believed that modern maize was derived from a wild species known to the Aztecs as *teocintli* (teosinte in modern parlance), which today occurs only in Mexico and Guatemala. However, breeding experiments have shown that, although teosinte and maize are both 10-chromosome species, they differ greatly in the sets of genes that each passes on from one generation to the next. It now seems likely that pod corns of the kind found in Bat Cave are the true ancestors of modern maize but that, at various stages in the evolutionary process, the maize cob acquired greater strength and rigidity through accidental hybridization with teosinte.

Maize plants and seeds vary greatly in size, appearance, and colour, but there are six main types.

In *pod corn* each kernel is wrapped in its own glumes. Although, as the probable ancestor of modern maize, it is of considerable interest to botanists, pod corn has little commercial importance. *Dent corn*, economically the most important maize, is so called because its kernels have small depressions in the crown. These dents are due to the differing rates at which the soft and hard starches in the kernel dry out during ripening. *Flint corn*, with hard, long kernels containing little soft starch, is widely grown in the colder soils of the northern corn belt, where it germinates more readily than dent corn, and also in tropical regions, where its hard seeds stand up better to attack by corn weevils.

Pop corn, familiar as a confectionery, is closely related to flint corn. Its value for popping derives from its small, very hard kernels, which are completely devoid of soft starch. The kernels are popped by heating, which causes the moisture in their cells to

Maize plant, from a botanical work of 1542—probably the earliest painting of the cereal to be published in Europe.

Both engravings on this page are based on paintings published in 1590 by John White, one of the early settlers in Virginia. Above: an Indian fans the flames beneath a cauldron containing fish and maize and other vegetables. Below: a corner of an Indian village, with two plots of ripening maize. The Indians of North, South, and Central America had grown maize for several thousand years before Columbus's time.

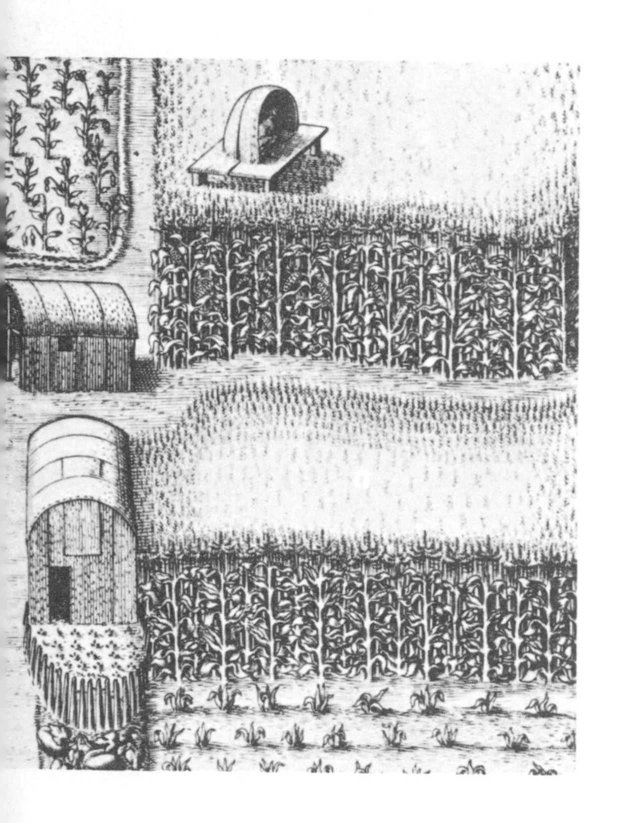

expand to the point at which the whole kernel explodes. Far from being a recent invention, popped corn was one of the earliest forms in which maize was eaten, since it is the simplest way to reduce the kernels to a soft, easily chewed food.

In contrast, *flour corn* kernels consist almost entirely of soft starch, which is commercially undesirable. Little used in the United States, flour corn is grown by many South American Indians in the mountainous regions of Peru, Bolivia, and Ecuador.

Finally, *sweet corn* is the type most widely grown for human consumption in North America and Europe. It owes its sweetness to the fact that only a small proportion of the sugars synthesized by the plant is converted into starch in the kernels.

Maize is cultivated in much the same way as wheat, but as the adult plant is much larger it needs considerably more space for development. The seeds are sown, at a depth of two inches, in groups of two or three at intervals of nine inches. The rows are usually about three feet apart. The weight of seed to the acre is about 10 pounds—only a tenth or less the rate for wheat. The average yield of grain, however, is about the same, from one to two tons an acre. In tropical regions maize is sown in the wet season to ripen in the dry one. In the temperate zone it is planted in the spring and harvested in the autumn. There are no winter corns: maize is frost-tender and freezing temperatures will kill it to the ground. As a general rule, maize can be grown only in regions where the farmer can count on at least 85 frost free days after planting. In North America and in some other regions maize is both harvested and "shelled" by machine.

The great bulk of the world maize crop—perhaps 90 per cent—is used as feed for livestock, including beef and dairy cattle, pigs, horses, and poultry. For cattle feed the whole plant is cut while still unripe and used as silage. Maize also has many industrial uses, mainly as a source of starch and sugars, including corn sugar (dextrose) and industrial alcohol; it is also used, as we have seen, in the production of whiskey.

As food for humans, maize is inferior to wheat and some other cereals because it contains very little protein. The absence of protein means that maize flour does not form gluten when it is mixed with water, and so maize bread is less palatable than wheat and rye loaves made from the "stretched" dough (full of air pockets) that gluten provides. Even so maize has formed the principal diet of many peoples in Central and South America for several thousand years. *Tortillas*, the daily bread of rural Mexico, are thin,

flat loaves made from maize flour. The North American *hominy* is one of several names for gruels made by adding water to coarse-ground corn meal. Manufactured foods, apart from pop corn, include cornflakes and other breakfast cereals. Finally, corn-on-the-cob (sweet corn) is an important vegetable in America and, to a lesser extent, in Europe.

World production of maize is about 230 million tons a year, of which about half is grown in the United States. Other major producers in the New World are Brazil, Mexico, and Argentina. In Europe, the principal producers are the Soviet Union, Romania, and Yugoslavia. Other countries growing several million tons a year include South Africa, India, and Indonesia.

Above: this machine can spray ten rows of maize at a time with insecticides, as here, or with chemicals to combat fungal and other diseases.

Right: in Mexico, maize farmers cover the developing cobs with bags to protect the seeds against birds.
Below: on large farms in the United States maize-harvesting machines have replaced the laborious job of cutting each individual cob by hand.

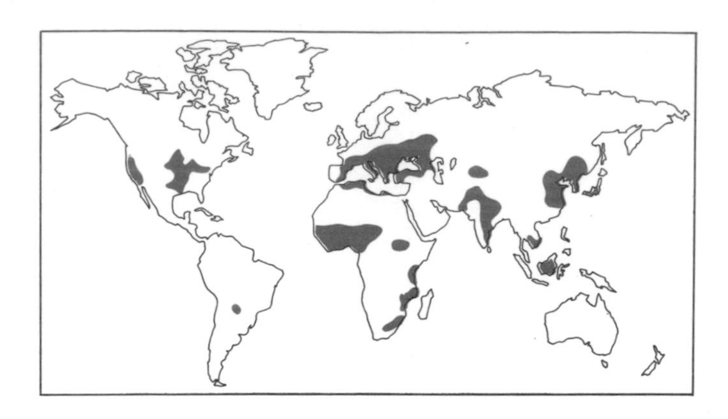

Map (above, left) shows the widespread cultivation of millet. As grain for human consumption, it is most important in the dry tropics of Africa and India. The plant (above) is topped by an ear of densely clustered, berry-like grains.

Millets

If maize is the great indigenous grain of America, the grasses known collectively as millets can be regarded as the leading grains of Africa. The most important millet belongs to the genus *Sorghum*. Known as *S. vulgare*, the species probably originated in Africa and was cultivated as a cereal in Egypt at least 4200 years ago. Today it is also grown in India, Pakistan, northern China and Manchuria, the United States, and, to a lesser extent, in the Soviet Union, the Near East, and Argentina. In India it is called *jowar*, in China *kaoliang*, in the Near East *durra*, and in the West Indies *guinea corn*—a reference to its African origin. In composition, the seeds of the varieties used as cereals are similar to those of maize, though they are somewhat higher in protein.

S. vulgare is a strong grass, its stem and leaves somewhat resembling those of maize. Its inflorescence, however, is completely unlike other cereals, for it develops small grains in dense clusters, the kernels varying in colour, according to type, from white to fawn, brown, or red. The flowers contain both male and female organs and are self-fertile, though a certain amount of cross-fertilization normally occurs in the wild. The discovery in 1952 of male sterility in some strains, however, has enabled breeders to develop hybrid seeds, which are now generally used by the larger growers in the United States and in some Asiatic countries.

The species was introduced from Africa to the United States (where it is called sorghum) in 1850. Normally, the plant grows to a height of from 6 to 16 feet. Recently, however, American breeders have succeeded in developing dwarf varieties that grow to about the same height as wheat and can be harvested by combine. In addition to grain sorghum, another variety of *S. vulgare*, called sweet sorghum, is used to make corn syrup and also for forage.

Below: this hog-millet on a Colorado farm is a low-growing variety that can be harvested by combine.

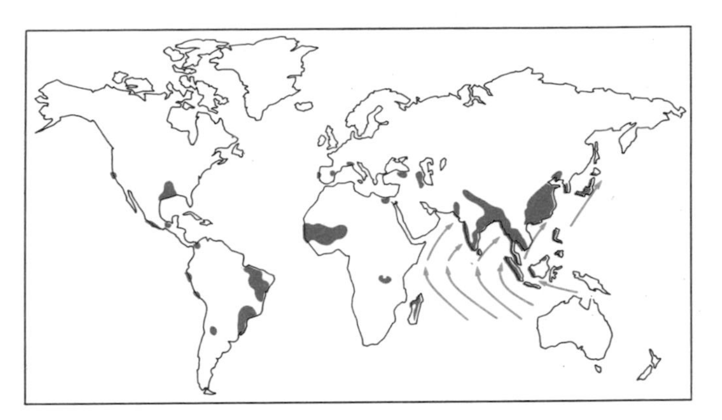

Map (above, right) shows main areas of rice cultivation and direction of the July monsoon. The delicate ears of rice (above) arise from stems that vary greatly in length according to variety.

Rice

Rice is the staple food (in some regions almost the only food) of considerably more than half the population of the world, and has been cultivated in southern and south-eastern Asia for many thousands of years. In India alone more than 8000 varieties of rice have been recorded; several thousand others have been developed in China, Burma, Thailand, the Philippines, and Japan. Almost all these varieties, however, belong to the single species *sativa* of the genus *Oryza*. The origins of this species are still obscure, though it clearly arose from wild marsh grasses, probably in India or China. Even evidence for the earliest date and location of paddy cultivation is vague. Chinese documents at least 5000 years old show that, at that time, the right to sow paddy was strictly reserved by the emperor or his nominees.

The swampy habitat of many of the wild species from which modern rice may have developed explains why rice (unlike any of the other major cereals) needs to grow in water for a large part of its life cycle. (Some varieties, it is true, are grown as "dry-land" crops in hilly regions, but by far the greater acreage

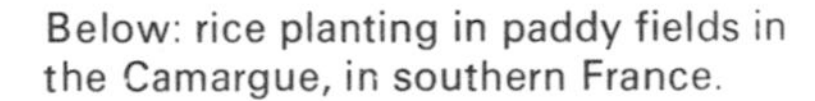

Below: rice planting in paddy fields in the Camargue, in southern France.

is planted in flooded soil.) For this reason, the principal rice-growing countries of the world lie in the monsoon belt, where heavy seasonal rains and simple but effective irrigation techniques enable the farmers to flood vast areas of soil to a depth of several feet.

The monsoons blow from the south-west or south-east from June to October, and from north-east or north-west between October and March. Only the June-to-October monsoon, however, can be counted upon to provide sufficient rain in all the Asian rice-growing countries; though twice-yearly harvests are possible in some regions.

In vast areas of Asia, cultivation, planting, harvesting, and maintenance of the irrigation systems is done on a communal basis. The first job of the season is the repairing and cleaning of the waterways and the dikes, or *bunds*, that determine the size and shape of the paddy fields. The bunds are made of compacted clay, thick mud, and weeds, and vary in height from place to place according to the level to which the flood waters commonly rise.

The soil of the paddy fields is turned over by ox-drawn plough or hand tools in order to bury the weeds. Ploughing and other preparations are usually done when the soil is under a few inches of water and so is easier to work. Often the soil is flattened by a spiked roller, which reduces lumps and "puddles" the mud.

Only in a minority of the rice-growing countries are seeds planted in the paddy fields. More usually, they are planted in nursery plots that have been specially prepared and enriched with manure. When the nursery soil has been thoroughly worked to a depth of about six inches, the water is drained off. The seeds are steeped in water for about 12 hours and are then kept in a cool place for a couple of days to allow them to germinate before planting. Nursery techniques differ widely from region to region, owing to variations in climate, soil, and local tradition. In cooler areas the seeds may be raised in "hotbeds" after soaking in warm water. In parts of western India the seed bed is covered with a layer of twigs underlaid by dried cow dung. These layers are then burnt, and the ash worked into the soil before the seeds are planted.

The seedlings remain in the nursery plot for periods ranging from 28 to 50 days, depending on the variety's growth rate and the condition of the paddy fields. The rate of seeding depends partly on the quality of the soil and partly on the growth rate but usually varies between 20 and 40 pounds to the acre. Experiments have shown that the timing of transplanting is highly

1

2

3

Above and right: 18th-century engravings, based on Chinese paintings, showing seven stages in rice cultivation and harvesting. (1) Flooded paddy is harrowed to mix the water and sun-baked earth; (2) the nursery-raised seedlings are planted out; (3) as the plants grow the level of the water in the paddy is raised; (4) the ripe harvest is gathered; (5) the sheaves are stacked; (6) the rice is threshed and, (7), ground into flour.

critical. If it is too early the shorter plants will be overwhelmed by the water; if too late, the yield will suffer. Seedlings suitable for transplanting are usually about seven to nine inches high, with five, six, or seven emerald-coloured leaves.

The nursery work and cultivation of the paddy fields are done concurrently so that the young seedlings can be transplanted when the soil is in the best possible condition to receive them and is free of weeds. The seedlings are planted in groups of two to six, each group from 4 to 16 inches apart. Like other cereals, the yield of rice grain depends greatly on the tillers developed by the young plant soon after transplanting. Tillering occurs at a specific stage in the life-cycle, and the development of fruitful tillers is to some extent determined by the spacing of the seedlings in the paddy field. If they are too close together the tillers will be poor and will produce seed either too late or not at all. If they are too far apart, the intervening patches of soil will encourage the growth of weeds, which will have the same effect on the tillers as close planting.

The plants grow with their topmost leaves and, later, their heads just above the surface of the water—the water level in the paddy field being adjusted by making gaps in the bunds. Obviously, however, it is a great advantage if the plant can withstand complete submergence in times of serious flood, and some varieties can survive up to 15 days of submergence as long as the water is not flowing too swiftly. The height of most varieties of the mature plant is between four and six feet. But in the valleys of some of the larger Asian rivers, such as the Me-kong, deep-water varieties attain a height of 20 feet and more.

Once the rice has flowered, the level of water in the paddy fields is lowered by shutting off the supply from the river and opening gaps in the bunds. As the seeds form, all the water is drained off, so encouraging the plant to build up its seeds at the expense of stem, leaves, and root. Eventually, the foliage withers and the crop is cut by hand.

The crop is threshed by spreading it out on mats and trampling it—often with the help of cattle. Then the husks are removed by winnowing; that is, by shaking the grain to and fro in a shallow, lipped tray made of plaited bamboo strips or palm leaves. The detached husks are blown away by the wind.

In contrast to these simple methods, rice growing in the wealthier countries is highly mechanized. In the United States, where rice is grown mainly in California, Louisiana, Texas, and Arkansas, machines are used at almost every stage. Even the levees (as the

Above: stages in the growth of the rice plant (water level shown in blue). About six weeks after sowing in the nursery plot, the seedlings are transplanted. The ears begin to develop about 10 weeks after transplanting, and the crop is fully ripe after another 8 to 10 weeks. The growing period from germination to harvesting averages about six months.

Left: a Japanese farmer sows rice seed in a nursery plot (top). Meanwhile, a nearby paddy field (centre) is ploughed to prepare it for planting out (bottom). Once the seedlings are established the water level will be raised almost to the tops of their leaves and will continue to be raised at intervals to keep pace with the growing crop.

Right: many rice-growing regions are so hilly that the paddy fields consist of tiny plots separated by a web-like network of bunds, as in the upper picture. By such means the farmer keeps the water level at the same depth in each plot. The fenced-off area is a nursery plot. When the rice crop has developed seed heads (lower picture), the water is drained from the paddy in order that each plant will concentrate its foodstuffs in the grain rather than in leaves, stems, and roots.

Indonesian rice (above) is tied into bundles before being air dried. The rice crop will rapidly deteriorate if it is stored when damp. Controlled air or sun drying reduces the water content of the grain to about 14 per cent.

The cakes (below), consisting of baked rice together with a little flavouring, are one of many forms in which Asian peoples use their basic cereal.

bunds are called locally) are built by mechanical means with a tractor dragging behind it two boards forming a wedge shape. (Earth levees may, in fact, be on their way out, following successful experiments with various types of plastic sheeting attached to stakes.) The soil is prepared in the usual way by ploughing and harrowing, and is then levelled with scrapers that remove the soil from high spots and deposit it in depressions. Many rice growers outside Asia do not raise seeds in nurseries but sow them directly in the paddy fields. In the United States sowing is often done from aircraft, which may also broadcast fertilizers at the same time. The crop is harvested by combine or reaper-binder.

Most rice eaten locally by the grower in Asia or exported to the wealthier countries is milled and polished to improve its appearance. Unfortunately these methods remove much that is nutritious in the grain because they get rid of both the outer layers and the embryo, and leave little behind except the starchy kernel. Among essential foodstuffs concentrated in the outer layers and embryo are proteins and small quantities of fats. But perhaps the most important loss is that of thiamine (vitamin B_1), which is essential to the body processes by which energy is released from carbohydrates. Since millions of people in Asia live almost solely on rice, they do not get thiamine from other sources. Its absence from their diet is directly responsible for the widespread occurrence of *beriberi*, an often-fatal disease that causes gross inflammation of the nerve tissues.

World production of rice is about 265 million tons a year, of which almost a third is produced by China and about a fifth by India. The other principal Asian producers are Pakistan, Japan, Indonesia, Thailand, Burma, and North and South Vietnam. Other large producers are Brazil, the United States, and Egypt.

At present, millions of people in Asia do not get enough to eat: malnutrition is their common lot, while starvation faces them on the all-too-frequent occasions of crop failure due either to drought or flood. The problem is one of both quality and quantity of food. As we have seen, for those peoples who live almost entirely on rice, even bumper crops provide an inadequate diet if the grain is stripped of much of its food value. One of the most important tasks now is to persuade people to clean and prepare rice in a much less drastic way than they do at present. Another suggestion is to use the flooded paddy lands not only for rice growing but also as breeding grounds for fresh-water fish, such as perch. This is not as fanciful

The Indian student (above) examining rice pollen is working for the U.N. Food and Agriculture Organization, which is engaged in a long-term programme to develop high-yielding rice varieties.

Below: rice grain. The endosperm is mainly starch. The embryo, scutellum, and outer layers of the grain contain protein and vitamins, but are usually removed in milling and polishing.

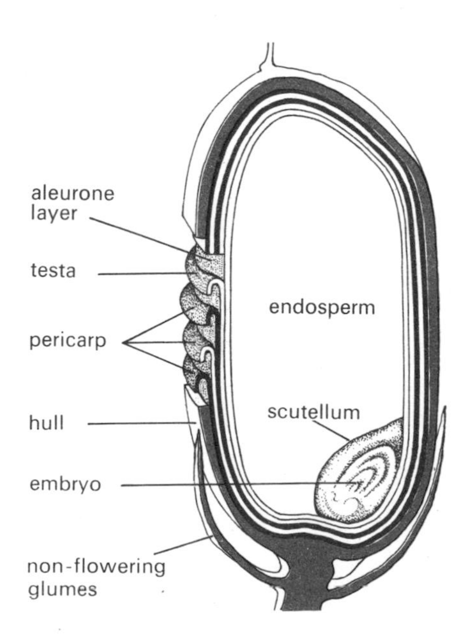

as it may seem, and it has been estimated that the flood plains of the greater river estuaries could provide a rich harvest of fish foods if the project was scientifically planned and run. Certainly, fish would provide a rich source of protein, which is seriously lacking in the diets of many continental Asians.

But the most serious problem concerns the perennial shortage of the staple food, rice. With the population of Asia increasing rapidly, there is only a limited amount of extra land that can be brought under cultivation. The essential need is greatly to increase the yield per acre. While in some areas of Australia, Italy, and Spain the yield is two or three tons an acre, in Asia yields of half a ton and less are common and yields of a ton or more are rare.

Mechanization is not necessarily the answer to the problem. Many of the traditional Asian methods of cultivation, though prodigal of labour (which is cheap anyway), are fundamentally sound. Moreover, much rice-growing in Asia is necessarily carried out on hilly or rolling countryside unsuitable for machines.

One of the most important improvements needed is in irrigation techniques, in order to conserve water in times of drought and to control it better in times of excessive flooding. In any given year, vast areas of south-eastern Asia can expect to lose as much as one fifth of the rice harvest owing to droughts or floods—or sometimes both. Another important improvement—and one that could be introduced quickly—is the development of better quality seeds by selection and hybridization. In some countries selection alone has boosted yields by 25 per cent. Hybridization, though a much slower process, could yield even better results in the long run.

In almost all the regions where the yield per acre is a ton or more, the rice crop is part of a rotation that includes other crops and pasture grasses. While there seems little doubt that rotation is at least partly responsible for these high yields, introducing rotation to new areas would pose some tricky problems. Not least would be the difficulty of persuading people to abandon their time-honoured custom of planting rice, year in, year out. It would be rather like telling the average European or North American to give up eating meat for a year or two. Indeed, it would be worse, since to many Asian peoples rice-growing is more than a means of providing food; it is a social activity upon which their very culture hinges. It is within a framework not only of desperate need but also of custom and culture that the battle for food in Asia must be fought.

4 Pasture Plants

Asked to name man's important food resources, most people would hardly think twice about pasture plants. Yet, after the grains, the grasses and clovers are our principle source of food. For although the human digestive system cannot deal with their fibrous stems and leaves, these plants form the principal diet of the cattle, sheep, and many other animals that provide us with meat and dairy produce. Since a single dairy cow needs at least an acre and a half of land for grazing and for winter feed, it follows that immense areas of the earth's surface need to be devoted to pasture and hay meadows. In the United States, for instance, no fewer than 1000 million acres (about three quarters of all the land available for agriculture) are classified as pasture—though much of it is on soil that would support little other than grasses.

As we saw in Chapter 2, the earliest herdsmen were nomadic, driving (or merely following) their herds, year in, year out, in search of adequate pasture. Even after they had domesticated the horse and the dog—so greatly increasing their mobility and their herding skill—they continued to allow their stock to roam freely over the available grasslands. This *free-range grazing* is still widely practised today, notably on the North American prairie, the South American pampas, the Eurasian steppe, and the South African veldt. On a smaller scale it is found on moorlands and in mountainous regions of Europe, where many such pastures have been communally owned by the villagers for centuries.

The disadvantage of free-range grazing is that the farmer has little control either over his stock or, more important, over the plants on which they feed. In this chapter we are concerned mainly with enclosed pastures, where the immense variety of available pasture plants—each with specific advantages in terms of nutritional value, growing habit, yield, and disease resistance—can be controlled by the good farmer to provide an altogether higher quality of grazing than is possible in most free-range pastures.

As fodder for cattle, sheep, and other animals, grasses and clovers are man's most important indirect source of food. The farmer uses his grassland in two distinct ways—as pasture for direct feeding (right) and as hay meadows whose crop is harvested (above, from a 14th-century manuscript) in spring and summer to provide winter feed.

Until the beginning of World War I few European farms were mechanized and most farmers had to rely on large numbers of temporary labourers to get in the grain and hay harvests. The photograph above, taken in 1905 on a Berkshire farm, shows 23 people and five wagons coping with the hay harvest of quite a small meadow.

Today the farmer has an enormous range of implements available to him. The finger-wheel rake (below) gathers the cut hay into rows and also turns it over to help it dry out before it is stacked.

Enclosed pastures are of two main kinds—*permanent* and *ley*. Permanent pasture is exactly what its name suggests; in many cases, it consists of a community of species that has developed, gradually and usually without any direct seeding on the part of the farmer, over a period of many years. Leys, on the other hand, are pastures that have been deliberately seeded for grazing or for the production of hay. On many farms, leys form part of a rotation, typically with wheat, root crops, and barley. Alternatively, leys may be produced on ploughed-up permanent pastures, grazed for several years, and then reploughed and seeded for grass again. Leys have been much in favour since the early years of World War II, when it was vital to get maximum yields from pastures and meadows. They undoubtedly provide a heavier yield than many permanent pastures, which until recent

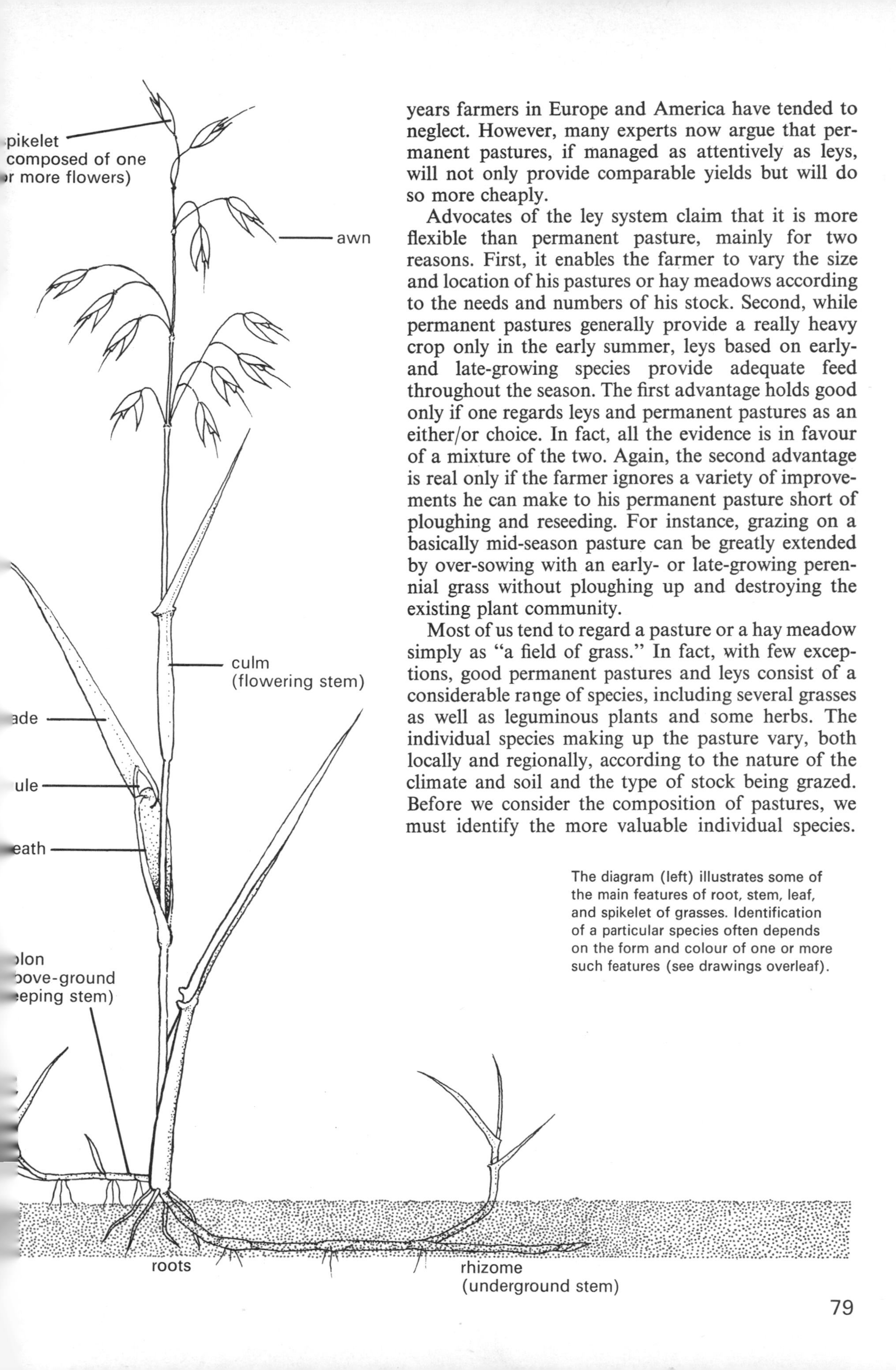

years farmers in Europe and America have tended to neglect. However, many experts now argue that permanent pastures, if managed as attentively as leys, will not only provide comparable yields but will do so more cheaply.

Advocates of the ley system claim that it is more flexible than permanent pasture, mainly for two reasons. First, it enables the farmer to vary the size and location of his pastures or hay meadows according to the needs and numbers of his stock. Second, while permanent pastures generally provide a really heavy crop only in the early summer, leys based on early- and late-growing species provide adequate feed throughout the season. The first advantage holds good only if one regards leys and permanent pastures as an either/or choice. In fact, all the evidence is in favour of a mixture of the two. Again, the second advantage is real only if the farmer ignores a variety of improvements he can make to his permanent pasture short of ploughing and reseeding. For instance, grazing on a basically mid-season pasture can be greatly extended by over-sowing with an early- or late-growing perennial grass without ploughing up and destroying the existing plant community.

Most of us tend to regard a pasture or a hay meadow simply as "a field of grass." In fact, with few exceptions, good permanent pastures and leys consist of a considerable range of species, including several grasses as well as leguminous plants and some herbs. The individual species making up the pasture vary, both locally and regionally, according to the nature of the climate and soil and the type of stock being grazed. Before we consider the composition of pastures, we must identify the more valuable individual species.

The diagram (left) illustrates some of the main features of root, stem, leaf, and spikelet of grasses. Identification of a particular species often depends on the form and colour of one or more such features (see drawings overleaf).

Grasses

From each of the half-dozen or so important pasture and meadow grasses, an enormous number of varieties has been developed in order to encourage particular properties in the plants. Some varieties, for instance, grow earlier or later than their wild ancestors; some tiller aggressively, others have had this tendency reduced to prevent them from crowding out other species; some withstand drought better than their wild-growing relatives; some recover quicker than others after intensive grazing or mowing. The important thing to remember about the development of varieties is that each of these qualities is intended not only to improve the individual species but also to complement the properties of other species with which they share the soil of a pasture.

Perennial ryegrass (*Lolium perenne*) is one of the most valuable grasses available to the stock farmer. It is almost invariably present in the best "natural" permanent pastures, especially on good loam and clay soils, and is also widely used in leys. Its main virtue is that it produces a thick bottom foliage of small, succulent leaves early in the spring while delaying the emergence of the seed head (which animals tend to ignore) until comparatively late in the season. Some varieties, especially those used with clover, have been bred to produce less tillers than the wild species, which tends to be too aggressive for some legumes. Others, developed for leys, give a longer seasonal period of leaf growth during the first two years.

The related Italian ryegrass (*Lolium italicum*) is a biennial and, though similar in appearance, is somewhat larger than the perennial. It shares most of the qualities of *L. perenne*, but in addition it remains palatable even when it has run to seed head. Italian is often used as a *catch crop*—that is, it is sown in spring after early potatoes have been harvested and is used for grazing or hay production until the autumn, when the soil is ploughed for a winter grain or other crop. Alternatively, it may be sown in the late summer to provide winter and early spring grazing for sheep.

Timothy (*Phleum pratense*), sometimes called catstail, is one of the best perennials on a wide range of fertile soils. A deep rooter, it produces a heavy crop of leaf during the early summer, when most other grasses are running to head. Although it does not grow vigorously during the autumn, such grazing as it does offer continues well into winter without loss of quality. While slow to establish itself in competition with more aggressive grasses, timothy is probably the most palatable of the perennials.

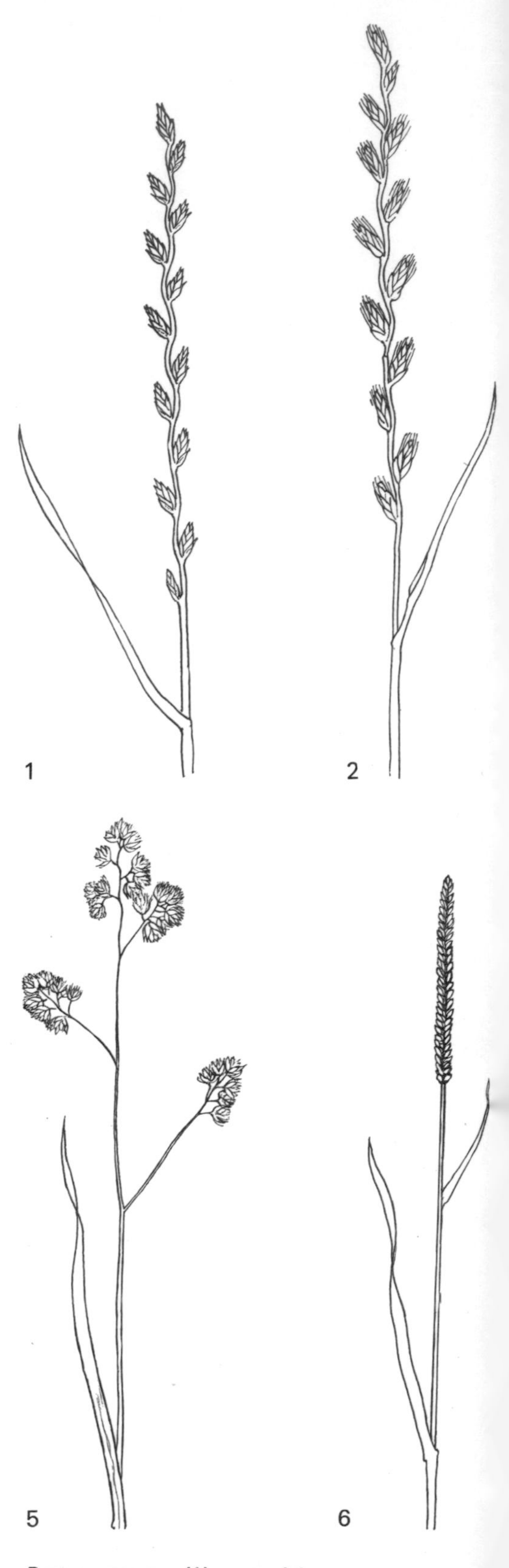

Pasture grasses: (1) perennial ryegrass, (2) Italian ryegrass, (3) timothy or catstail, (4) meadow fescue, (5) cocksfoot, (6) crested dogstail, (7) buffalo grass, (8) Kentucky bluegrass.

Meadow fescue (*Festuca pratensis*), another deep-rooting perennial, is an ideal companion for timothy, especially when both are grown with clover in grazing leys, because it provides a good, heavy growth of leaf during the spring. It also has a good *aftermath*, or second growth after cutting.

Cocksfoot (*Dactylis glomerata*) is the deepest rooting of the better permanent grasses and so is especially valuable in pastures liable to summer drought. Some farmers regard cocksfoot as less palatable than the other perennials because it becomes woody with approaching ripeness. For this reason it should be kept short by close grazing or early cutting. This also helps the growth of clover, which otherwise may be crowded out.

Crested dogstail (*Cynosurus cristatus*) is found in many pastures on poorer, lighter soils. It is low growing (too low for mowing) but it is readily eaten by sheep. Dogstail is a poor yielder but, by occupying the spaces between larger perennials, helps to keep pastures free from undesirable self-sown, or *volunteer*, grasses.

Smooth meadow grass (*Poa pratensis*), also found in many permanent pastures, is better known by its American name, Kentucky bluegrass. Other species, not highly regarded on enclosed European pastures but widespread and valuable on the American prairies and elsewhere, include bent or red top (*Agrostis alba*), couch grass (*Triticum repens*), and buffalo grass (*Sesleria dactyloides*).

Some grasses develop roots that are considerably longer than their stems. The eroded bank (above) on a farm in Arizona shows blue grama grass roots extending to a depth of four feet.

Leguminous Plants

Today, many of the best leys and permanent pastures comprise a mixture of grasses and legumes, usually clovers. Legumes are important because of their capacity to "fix" atmospheric nitrogen present in the soil with the help of bacteria living in nodules on their roots. Hence, the legumes are the chief source of protein in pastures, and complement the predominantly carbohydrate diet offered by the grasses. But the value of legumes does not end there. It has been shown that a pasture containing a mixture of ryegrass and white clover is not only twice as productive as similar soil containing ryegrass alone, but that the yield of ryegrass is 50 per cent greater in the mixed pasture. The reason is that the ryegrass is able to exploit for its own growth the nitrogen that passes into the soil from the root nodules of the clover. Legumes, then, enrich the soil as well as provide protein for livestock. For this reason, clover or mixed clover and grass leys are a valuable part of a rotation, and are often sown on arable land immediately before cereals.

True clovers belong to the genus *Trifolium.* White clover (*T. repens*) is the main indigenous clover in British permanent pastures, and has been one of the most important perennials on fertile grazing lands for many hundreds of years. There are two main types—a low-growing, creeping clover, and a somewhat taller type with long petioles. The best varieties of both can usually hold their own with competing grasses,

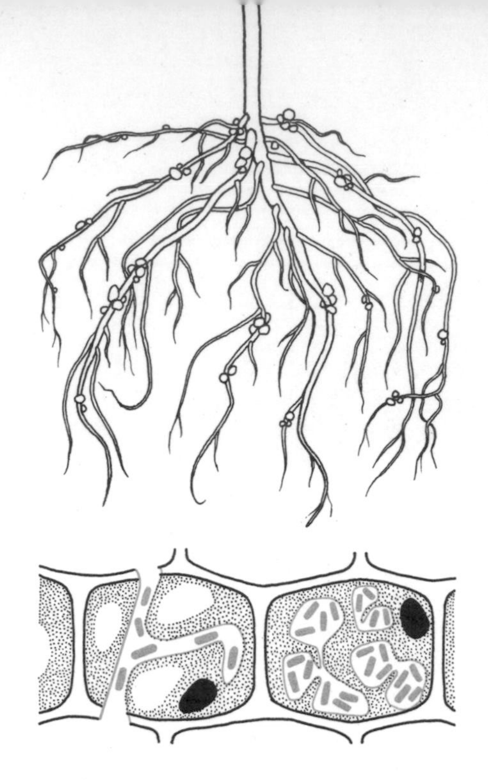

Above: legumes have root nodules (upper picture) whose cells contain nitrogen. Nodules form when the root cells are invaded, and stimulated to divide, by nitrogen-fixing bacteria (lower picture).

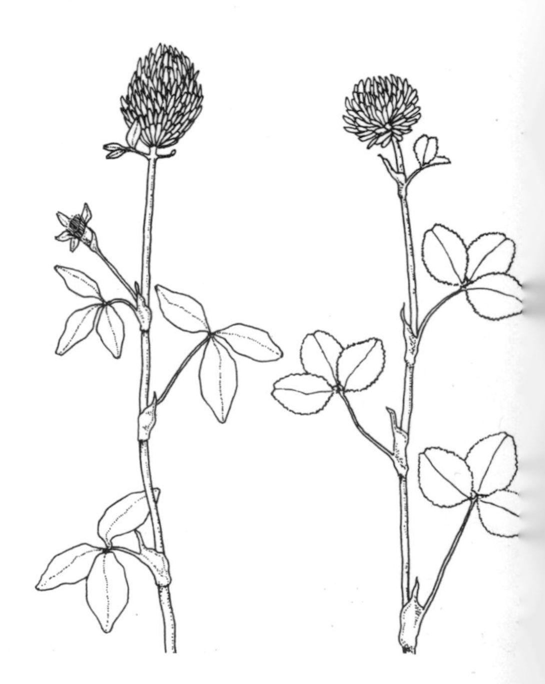

Below (from left to right): white, broad-red, and alsike clovers.

The two pasture legumes above, lucerne (left) and sainfoin, have longer, more-open flower heads than most clovers.

Below: ryegrass and white clover form the basis of a highly productive mixed pasture, the ryegrass being stimulated by the legume to develop higher yields than it could achieve on its own. The deliberate sowing of pastures of these two species began in the 17th century.

and the creeping type will stand up to continuous hard grazing by sheep.

Broad red clover (*T. pratense*) is commonly mixed with Italian ryegrass in one-year leys that are to be used for hay rather than grazing. It produces a strong aftermath that may also be mown, or may be grazed or taken for seed. There is also a late-flowering red clover, used mainly for grazing, that will continue to flourish in a ley for two or three years.

Crimson or scarlet clover (*T. incarnatum*), known to many farmers simply as "trifolium," is a true annual. It is often used (alone or with Italian ryegrass) as a catch crop between cereals and roots, when it may be cut or intensively grazed in May or June. Though one of the prettiest of clovers, it is less palatable than the others.

Alsike clover (*T. hybridum*), with red and white flowers, is very common in North America. Compared with white clover, it is a poor yielder on good pasture soils, but it is valuable on damp, acidic soils where other clovers normally fail.

Of legumes other than clovers, lucerne or alfalfa (*Medicago sativa*) is by far the most important, and is the major forage crop in many hot, dry countries. One reason for this is that lucerne is extremely deep rooting and so is strongly resistant to drought—a property that is also important during periodic dry spells in the temperate zones of Europe and North America. Lucerne is one of the most productive pasture plants, often providing four crops of hay a year and giving a dry-matter yield of as much as 8000 pounds per acre—some 25 per cent more than the farmer can expect from a grass-clover ley. It has high nutritional value, especially in protein, and readily lends itself to hay making since it is normally grown on dry soils. Lucerne is usually planted with a companion grass, preferably timothy or meadow fescue; ryegrass and cocksfoot, however, are too aggressive for it. Hitherto somewhat ignored in Britain, lucerne is now making headway on dairy farms.

Other legumes include sainfoin (*Onobrychis sativa*), a perennial that gives good yields for three or four years and is extremely palatable; and trefoil (*Medicago lupulina*), a useful annual on chalky soils, where it is often sown with a cereal to provide a stubble feed for sheep after the harvest.

Many farmers like to see small quantities of herbs in their leys. These have little value in terms of nutrition or yield, but may help to crowd out less desirable plants. Among the most popular of such herbs are chicory, plantain, field parsley, and burnet.

Seeding and Management

As we have seen, both permanent pastures and leys almost invariably consist of a mixture of species. There are many advantages in a mixed plant community. Different species tend to produce their maximum yield at different times of the year, so a well-chosen mixture will provide a good "bite" throughout the season. Moreover, different species are resistant to different diseases, so that if one member of the community succumbs to a particular disease, the chances are that the other plants will continue to provide stock with enough food. At one time pasture seed mixtures often contained a very large number of species. Today, however, we know a good deal more about the properties of individual varieties and about the compatibility of mixtures, and most leys contain no more than five or six planted species, together with a number of volunteers—more or less desirable species that inevitably invade a pasture from a neighbouring field or hedgerow. Mixtures vary, of course, according to climate, soil, and the particular stock to be fed. But a typical mixture will comprise (for each acre of pasture) five or six pounds each of ryegrass, timothy, meadow fescue, and cocksfoot seeds, and two pounds of white clover.

The variety of each species selected depends on many factors. The aim is to provide as heavy a crop as possible of palatable food throughout the spring, summer, and autumn; but the best grazing varieties are not necessarily the best for hay or silage.

Electric fences, such as the one above, are a simple method of confining cattle or other stock to one particular area of a pasture. The wire, which gives a mild electric shock, can be moved from one part of the pasture to another.

Below: the diagram shows synchronization of pasture growth (green) and the food needs of a dairy herd (black). Maximum pasture is available to cows in calf about mid-May. Fall in grass production in July coincides with fall in appetite (and milk production) of the herd. In the early autumn late-season grasses provide feed for cows and young calves.

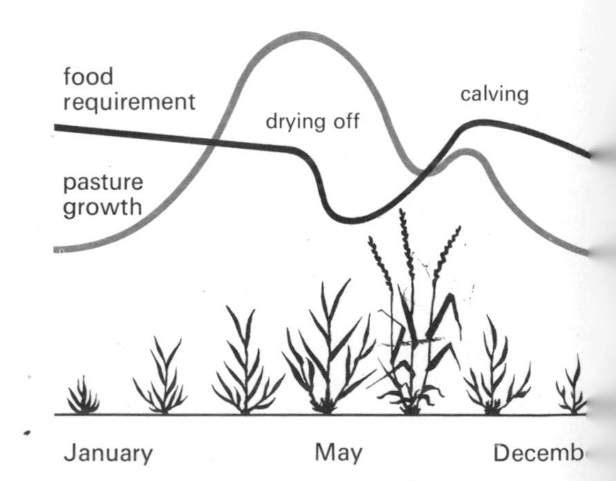

Fertilizers are as important to the management of pasture and meadow as they are to the raising of any other crops. The assumption (erroneous, as we have seen) that leys are invariably more productive than permanent pastures is due, as much as anything, to the fact that many farmers do not realize that the latter respond to fertilizers just as readily as do leys. Every crop removes a proportion of the nutrients available in the soil. In a pasture, some of this is returned in the form of faeces and urine from the stock grazing on it. A meadow set aside for hay or silage, however, is denied this source of replenishment, so the wise farmer sees to it that the areas of pasture reserved for cutting change during the course of each season. Often, for instance, a hay meadow will be used for grazing after the first cut in May or June.

In spite of this, the soil of pasture and meadow suffers depletion from time to time, and the nature and extent of depletion can be exactly determined only by analysing soil samples every five years or so. The four basic fertilizers, as we have seen, are lime, potash, phosphate, and nitrogen. Lime neutralizes soil acidity and so is especially important to the growth of legumes. But excessive use of lime seems to inhibit the capacity of legumes to take up phosphates in the soil—and it is through the generous use of phosphate that the farmer relies on a heavy yield of clover. Analysis of the soil helps the farmer to strike a balance in his use of lime and phosphate.

The decision to use nitrogen (either as nitrates or as ammonium sulphate) on pastures is critical, for it involves more than simply increasing crop yields. Nitrogen encourages grasses to grow aggressively and, thus, to crowd out or actually to suppress the legumes, which are themselves a rich source of nitrogen. In short, the farmer has to decide whether to grow clovers or whether to rely on the rich growth of grasses possible with applied nitrogen. Experiments have shown it is pointless to compromise: light applications of nitrogen tend to discourage the legumes without promoting a continuous and steady rise in grass production. The farmer's decision depends partly on economics and partly on the nature of his stock. A mixed clover/grass pasture is cheaper to establish and maintain than a pure grass pasture treated with applied nitrogen. But even the best clover/grass pastures cannot compete with the sheer bulk of highly nutritious feed yielded by good grasses that have received heavy dressings of nitrogen.

Can the farmer translate these extra-heavy crop yields into profits? The answer is yes—provided he

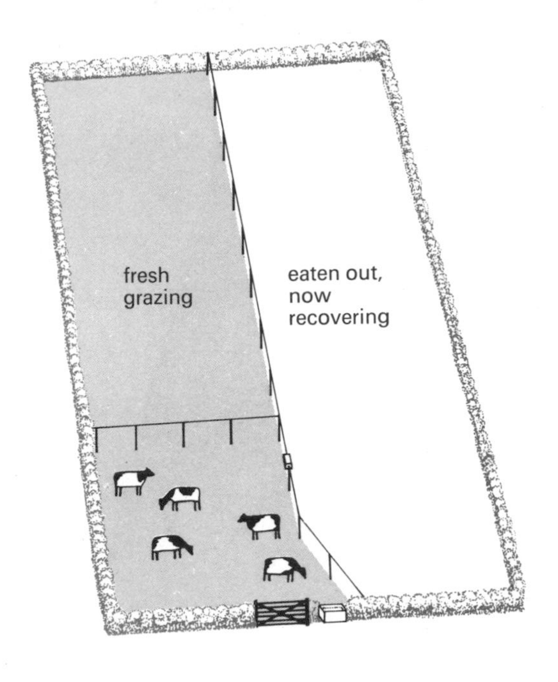

The diagrams show two ways in which a permanent pasture or ley can be divided up into folds for cattle (above) and sheep (below). The cattle pasture is split lengthwise by a semi-permanent electric fence, while movable electric fences confine the stock to a specific fold. Note that in the sheep pasture the lambs take first bite in each fold, which is then eaten out by the ewes. Close control over the stock, as in these two examples, enables the farmer to avoid overgrazing, which can damage the pasture in spring, or undergrazing, which in autumn encourages growth of poor grasses at the expense of clover.

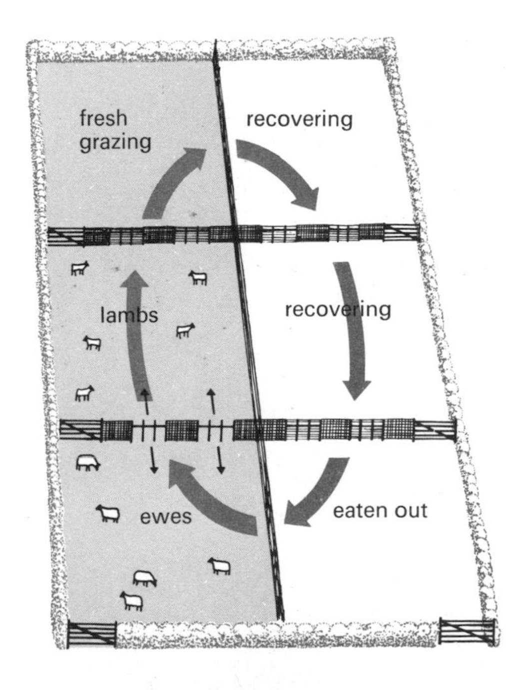

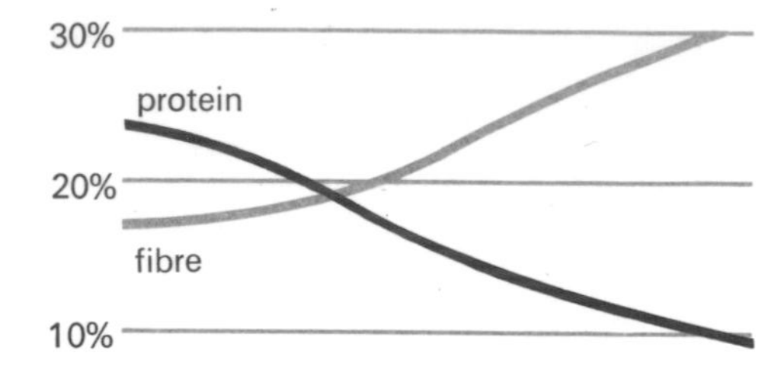

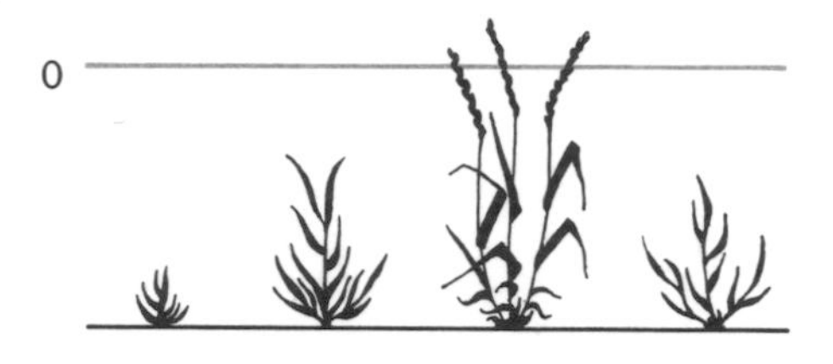

Graph (above) shows how food value (as protein) falls as fibre content rises in spring grasses. The curves intersect in mid-April. Grass for dairy-cattle silage should be cut before May. The pasture will then yield a second cut, or provide grazing, later in the year.

Below: forage harvester cutting silage grass, which need not be field-dried.

intensively grazes his pasture with stock in which high production gives high returns. In Britain (though not everywhere else) this means dairy cattle: nitrogen-fed pastures can increase milk yields to the point where the additional profits more than make good the extra expense of the fertilizer. For beef cattle and sheep, where production increases are less dramatic and where, in any case, the profit margins are smaller, clover/grass pastures are a better bet.

Winter Feed

Under ideal conditions, stock would be able to graze on highly productive pastures throughout the year. Unfortunately, most pasture plants do not grow throughout the year in any given region; even the winter varieties that have been developed do not provide sufficient nourishment to enable dairy cattle, for instance, to maintain their output of milk during the winter in temperate climates. In the savannah grasslands of semi-tropical regions, especially in parts of Africa, there is a lengthy pause in the growth of pasture plants during the dry season, when there may be a complete absence of rain for many weeks at a time.

Hence the stock farmer must set aside a proportion of his pasture crop for use as winter, or dry-season, feed in the form of hay or silage. The best hay and silage are produced from pasture plants that have not developed seeds. This is because, once the seeds are formed, they monopolize most of a plant's food supplies, and the stems and leaves (which are the most palatable) become woody and less digestible. This does not mean that hay and silage crops are always cut in the spring or early summer. As we have seen, many grasses and clovers, as well as lucerne, provide more than one cut a season, while several varieties have been especially developed for late growth.

Before hay- or silage-making, the crop must carry as little free moisture (i.e. rain water and dew) as possible; indeed, in the case of hay, most of the natural moisture within the plant cells must also be removed. In new-mown hay, water contributes anything up to 80 per cent of the total weight of the plant, and this must be reduced to about 14 per cent if the product is to provide a satisfactory feed. If stacked when damp, hay may ignite spontaneously, owing to the growth of countless millions of bacteria and moulds in the interior of the stack. So, after it is cut, grass for hay is left in the fields for a week or two to enable it to dry out; usually, it is turned over or tossed several times by machine or by hand so that sun and

Above: silos on a Northumberland farm.

Below: a farmer controls the flow of fine-cut grass into his silo. The small flakes of grass must pack down tightly, excluding air, so that fermentation can take place with the help of bacteria imported on the leaf.

wind can get to every part of the crop. In countries such as Britain, where the summer weather is highly changeable, haymaking is a gamble. The farmer often faces the choice of cutting his meadows a little too early or too late in the hope of taking advantage of spells of dry weather that may, in any case, turn to rain before the hay has had time to dry.

Ensiling has an advantage over haymaking in that the grass does not have to be dried in the field. The crop is cut with a grass harvester, sliced into small pieces, and then deposited in a pit or, better, a *silo*—a tall, circular tower in which the grass undergoes a natural pickling process. The point of reducing the plants to small pieces is to enable them to pack down into a dense mass containing no air pockets: the absence of air is essential to the pickling process that transforms the plants into silage.

Pickling is carried out by bacteria that were already present on the plants in the meadow. Under pressure and in airless conditions, enzymes in the bacteria carry out a process of fermentation. As we know, fermentation depends on the presence of sugars. These are, of course, abundant in grasses; but the farmer often accelerates the fermentation process by spraying molasses (a derivative of sugar) onto the plants as they are packing down in the silo. The use of molasses is essential with silage made from clovers or lucerne, which have a lower concentration of sugars than grasses. The fermentation process gives rise to a large number of products, one of the most important of which is lactic acid, a highly nutritious food that is also present in sour milk.

During and since World War II there has been a tendency for farmers to concentrate on silage production at the expense of hay. If the farmer has the necessary equipment for quick harvesting, ensiling certainly offers a more reliable method of securing good winter feed, especially in regions with rainy summers. But the claim of many farmers that silage offers a more nutritious feed, especially for dairy cattle, holds only if one compares good silage with the somewhat indifferent quality of hay that is all too common in both Europe and America. Experiments in both Britain and the United States have shown that high-quality hay is just as nutritious, if not more so, than high-quality silage. In any case, hay versus silage is a somewhat sterile field for argument. Ideally, the farmer should provide both good silage and good hay, since most stock respond well to periodic variations in their grass diet, in addition to alternative feeds, such as cattle cakes and root crops.

5 Sugar Cane

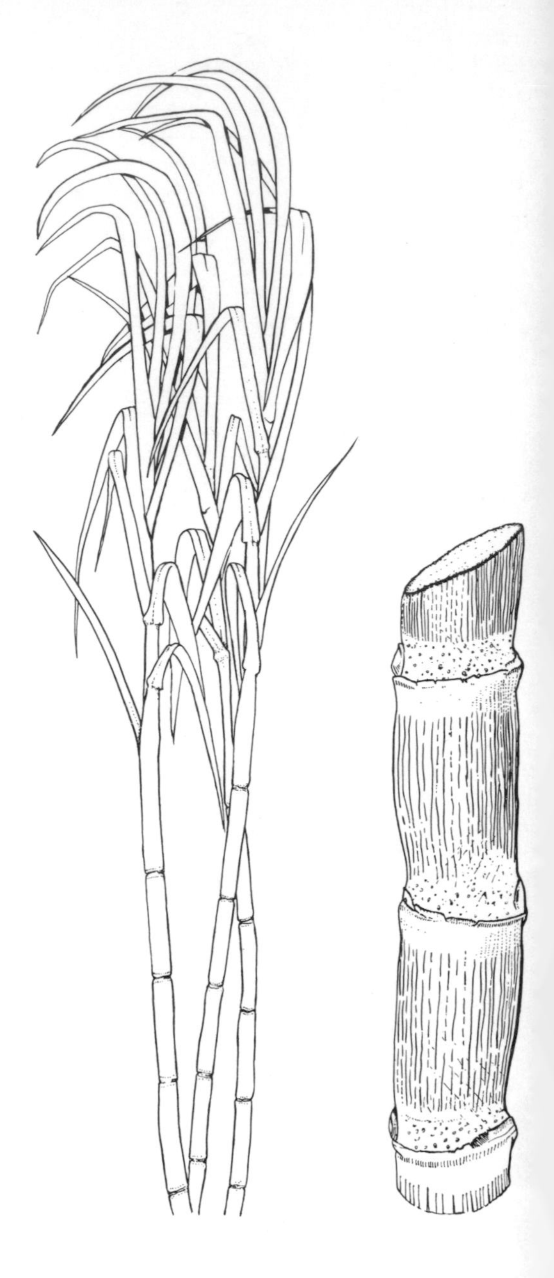

The cereals (Chapter 3) and the main pasture plants (Chapter 4) all belong to the family of grasses. But, as we have seen, we use cereals in an entirely different way from pasture plants: we cultivate the former for their seeds, whereas we cultivate the latter for their stems and leaves. This means that we must harvest the two types of grasses at different stages in their life cycles. Grass stems lose most of their food value when they begin to develop seed: as hay they are then useless; as pasture they are often ignored by cattle.

There is another grass—a giant—that man harvests, like hay, before it runs to seed: this is sugar cane (*Saccharum officinarum*). It is the biggest of all grasses economically important to man, growing to a height of 10 to 20 feet and sometimes more. The species originated in Asia and is probably due to hybridization among its wild relatives, *S. spontaneum* (found in India, the Philippines, Malaysia, and Indonesia) and *S. robustum* (found mainly in New Guinea). However, neither of these wild species contains as much sugar (sucrose) as *S. officinarum*, the modern varieties of which have a very long history of selection and cross-breeding by man.

Sugar cane is easy to grow. It needs a rich, deep, well-drained soil, and a hot, moist climate (with temperatures averaging about 70°F, and an annual

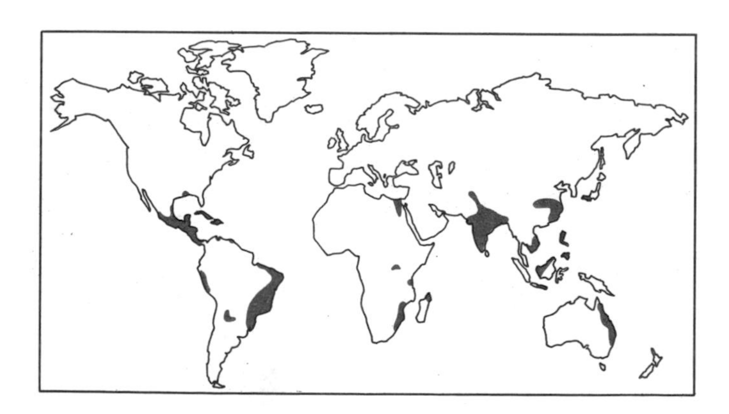

The sugar cane plant (above, left) is one of the largest members of the grass family. The plant stores sugar mainly in the stem. New plants are raised from seed pieces (above, right) cut from the stem, new buds arising from the nodes.

Map (left) shows main areas of sugar-cane cultivation. Although the plant is indigenous to Asia, the principal areas of production today are in the New World, where it was introduced by Columbus. Cane provides three fifths of the world's sugar, the remainder being extracted from sugar beet (see page 98).

De Bry's engraving of 1599 (above) shows the gathering of the cane harvest, the crushing of the cane, and the boiling of the sap into a thick syrup, on a North American plantation. The workers are supposed to be African slaves. The grimmer realities of slavery in Jamaica are shown (right) in this 19th-century engraving from an anti-slavery pamphlet. The slave wore a head frame to prevent him from eating the cane sugar; heavy weights hanging from a chain attached to a belt locked around his waist hampered any attempts to escape.

rainfall of around 60 inches, which may be supplemented by irrigation). Although the cane stalks consist mainly of cellulose and stored sugar—both the products of photosynthesis—the plant removes considerable quantities of nutrients from the soil, which must be replenished regularly with dressings of nitrogen, potash, and phosphate fertilizers. Like maize, sugar cane cannot withstand competition from weeds, and the soil must be well cultivated before and after planting.

The crop is raised from stalk (stem) cuttings, or *seed pieces*, between one and two feet long and containing two or more of the well-defined joints, or nodes. The cuttings are placed end to end in rows four to six feet apart and covered by about four inches of soil. Buds at the nodes sprout within a few days of planting, and rapidly develop roots and primary shoots and tillers. The crop is allowed to grow for a period that varies, depending on the region, from 8 to as much as 30 months, though 11 to 16 months is the average in the main producer countries. The cane is usually harvested toward the end of the rainy season, when the leaves and stems are rich in sugar but the plants have not yet developed seeds. At this point the stalks, which may be yellow, green, red, or purple in colour, are between one inch and three inches in diameter.

West Indian plantation workers (above) sort and cut cane into seed pieces for planting. Plants from these cuttings will yield up to five harvests; the soil is then ploughed up and replanted. The planter (below) places the pieces end to end in a furrow. In the West Indies the cane usually produces its first crop about a year after planting.

On most sugar plantations the cane is still cut by hand with a machete (below), though in Australia and Hawaii, where labour costs are high, mechanical cutters are used. The cane, cut into four-foot lengths, is loaded onto cage-like trucks (above) and taken to the factory. The unused cane leaves are used as a cattle feed or as fertilizer.

Sugar cane is a perennial. After it has been cut, usually a few inches above the surface of the soil, the stubble develops new shoots and roots that will, in due time, provide the next crop. Two or three of these *ratoon* crops, as they are called, are usually possible before the soil is ploughed up and new stalk cuttings planted. Normal crop yields are about 50 tons of cane to the acre, from which about five tons of sugar can be extracted.

Although in Australia and parts of the United States much of the crop is harvested by machine, the major portion of the world crop is still cut by hand with the help of a machete or other heavy, sharp knife. This calls for an abundant supply of cheap labour. During the 18th and early 19th centuries the production of sugar in the Caribbean islands and in the southern United States depended entirely on slaves imported from Africa. British slaves were emancipated in 1834, and within a few years sugar production in many of the smaller islands of the West Indies had either slumped greatly or ceased altogether. Even today, the profitability of most of the major sugar plantations depends on low-paid manual labourers, who are often imported from underdeveloped countries with high unemployment. The future will almost certainly see an increase in mechanical methods of harvesting, in spite of their high capital cost.

Above: cane passing into the crushing mill. After the sugar-containing tissues have been removed, the remaining fibres, called bagasse, are used as fuel to fire the boilers at the mill.

The cut cane is transported to the sugar mill, where it passes through mechanical crushers that break down the fibres and release the sugary sap. In many mills the fuel that powers the crushers consists of dry, waste fibres, called *bagasse*. This fuel also provides heat for boiling the sap, so as to concentrate it, first into a thick syrup and then into a solid mass of brown crystals. The familiar white sugar to which we are accustomed is due to further refining of the brown crystals, which are re-dissolved and treated with various chemicals that bleach the crystals and remove certain organic matter. But there is also a market for the crude sugar in the form of a dark syrup called *molasses*. This is used to raise the carbohydrate content of silage (see p. 87); it is also the basic ingredient in the production of rum, by fermentation (which is the reason why the main sugar-producing countries are also the main rum-distillers); and it is used in many industrial processes.

Fermentation is an ever-present problem in sugar refining. Just as grapes carry with them from the vineyard the yeasts that ferment grape sugar into wine, so crushed sugar cane carries with it enough wild yeast to ferment the sugar to alcohol, which is then converted to vinegar by the action of a bacillus, *Bacterium aceti*. This bacillus constantly lies in wait, as it were, to attack the products of fermentation, so

Below: production of raw sugar from cane. Juice from cane is mixed with lime and heated to concentrate impurities, which are removed in the clarifier. The clear juice, concentrated in evaporators, is boiled in the vacuum pan, emerging as a mixture of crystals (which are refined later) and molasses.

crusher
cane
roller
bagasse
juice meter
lime and water
juice liming tank
heater
heater
clarifier
scum tank
sweet tank
filter press
mud to fields
vapour
moisture evaporators
vapour
vacuum pan
mixer
spinner
weigher
raw sugar to store

Above: raw sugar awaits refining in one of the largest stores of its kind in the world. Most raw sugar is refined in the countries in which the finished product is ultimately sold.

the crushing and refining of the cane sap must be carried out quickly if trouble is to be avoided. As soon as the syrup has been concentrated by boiling or, as in more modern refineries, by evaporation under reduced pressure, it becomes immune to attack either by yeast or by the vinegar bacillus. (This is why jams and jellies, containing a high concentration of added sugar, will keep indefinitely.)

The need for immediate crushing and refining at the mill creates, in turn, a harvesting problem. If the entire crop of a given area arrived at the mill at the same time, a considerable proportion of it would lie idle for several days before processing. Unless cane goes through the crushers within two days of cutting, its sugar content drops, just as hay loses much of its food value if it lies too long in the meadow. The problem is solved by planting (and hence harvesting) the various cane fields on each plantation at different times; the interval between first and last plantings may be several months.

World production of sugar is at present about 53 million tons; of this, about 32 million tons is derived from sugar cane, the remainder from sugar beet (see p. 98). The principal cane-growing countries are Cuba (about 4 million tons), Brazil (more than 3 million), and India (more than 2 million).

Below: sugar refining. The centrifuge removes the film of molasses sticking to the crystals, which are liquified and purified with lime and carbon dioxide in a pressure filter. The amber liquid is made colourless in the charcoal filter and is recrystallized as pure white sugar in the vacuum pan.

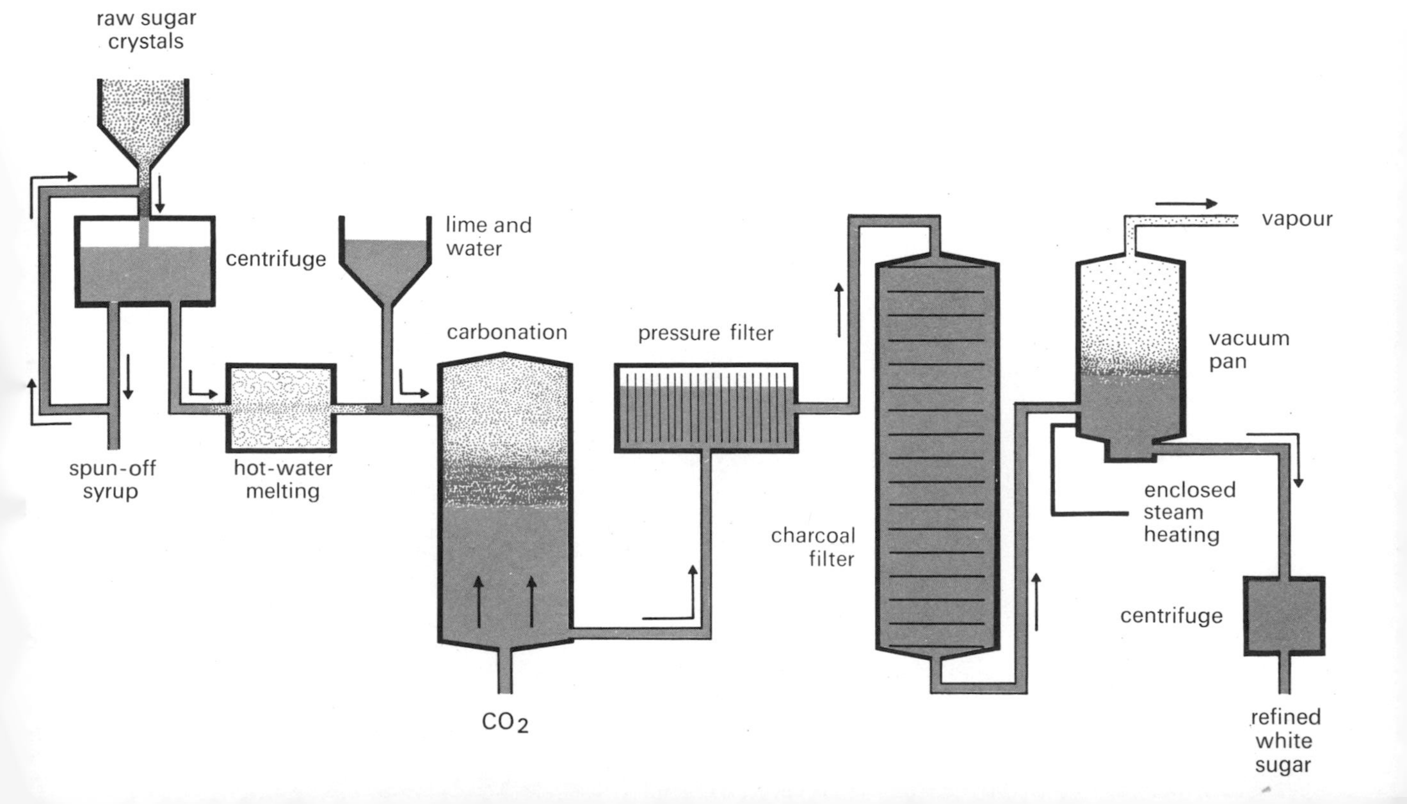

6 Roots, Bulbs, and Tubers

Most of the higher plants reproduce themselves in one of two ways. The first, and more familiar, is the sexual process in which cells produced by a plant's male and female organs fuse to form a seed. This process is analogous to the production of the foetus in higher animals, including man, and is common to the great majority of plants. Many, however, are able to propagate by *vegetative reproduction*—that is, by growing new stems, roots, and shoots that develop into a new plant or an extension of the old one. In a wide range of species, this vegetative process is additional to normal sexual reproduction, but in some rare cases it takes its place.

The seed of a plant, as we have seen, contains a rich store of food. But its survival in nature involves many hazards, of which drought, cold, and the appetites of birds and other animals are among the most important. In many ways vegetative reproduction offers a safer alternative, since it usually takes place beneath the surface of the soil. Even so, the plant still faces the problem of survival in cold or dry seasons. The species we consider in this chapter deal with this problem by developing underground food stores—usually in the form of *rhizomes* (creeping stems) or *tubers* (swollen masses that resemble roots but are in fact underground

Potato (above) reproduces vegetatively, buds arising from the "eyes" of the tuber. Stages in this process are shown below: tuber develops shoot and small roots; new tubers develop in summer, when plant is in full leaf; in autumn the haulm (stem and leaves) withers, the plant storing food in the tubers.

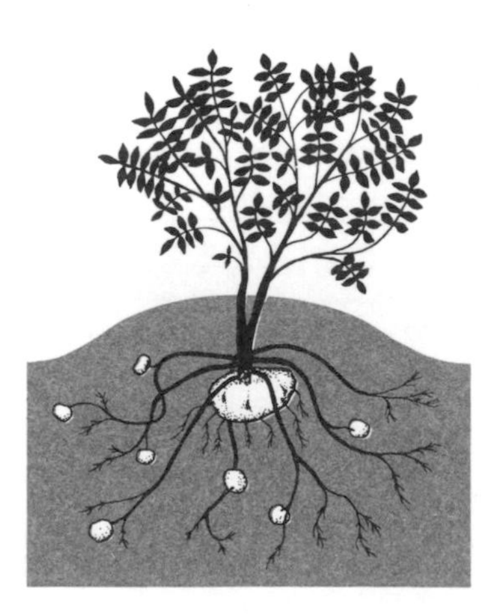

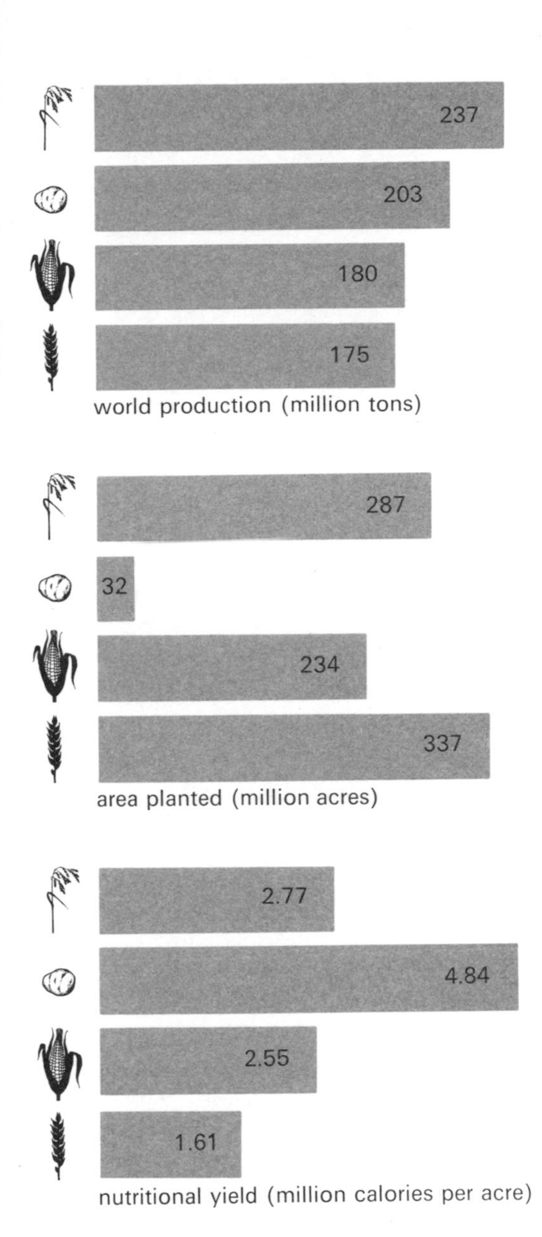

The diagrams above compare production, acreage planted, and nutritional yield of rice, potatoes, maize, and wheat. Although a less-concentrated food than the others, potatoes require much less acreage, and so their nutritional yield per acre is the highest of the four.

extensions of the stem); we shall also be considering other species that, although reproduced sexually, have also adapted themselves to survive unfavourable seasons of the year by developing enlarged roots, or *bulbs* (modified stems) that are rich in the foodstuffs that the plants need for survival.

These plants are mainly biennials or perennials. The biennials build up their food reserves during their first year in order to survive until the following spring, when flowers and seeds develop on their new-season shoots. Man interrupts this two-year cycle by harvesting the plants before or during their first winter. But since he can raise such plants only from seed, he must leave a proportion of his crop in the ground until they have developed seeds in the second year.

With perennials, however, the farmer does not need seeds (unless he wishes to develop new varieties): he starts with a root or tuber, which sends up a shoot and bears flowers and foliage in its first growing season. The plant uses its food reserves to form fresh roots or tubers, so that when the farmer harvests his crop each original root or tuber will have disappeared and a large number of new ones will have developed in its place. The farmer takes most of these for food and uses the remainder to produce the next year's crop.

Although of immense value to man, these plants offer a less concentrated form of food than grains (compare the food values for wheat and the potato on page 15). This is because roots and similar crops store a great deal of water. In simple terms, two pounds of wheat have as much food value as seven pounds of potatoes. Moreover, owing to their high water content these crops are bulky to store, heavy to transport, and difficult to keep in good condition. Whereas grain will keep indefinitely as long as it is dry, the root crops must be kept cool and moist and protected against frost; even then, they will keep through only one resting season—a maximum of nine months.

It may seem surprising that we should bother to raise crops with such obvious drawbacks. In fact, they have many advantages too. They are easy to grow and have exceptionally heavy yields compared with grains; for this reason they are well suited to small-plot cultivation, typical of peasant farming in many regions. Moreover, many roots provide an excellent fodder for cattle, sheep, and pigs. Finally, these plants have an important role in the rotation of crops, because they demand deep and heavy cultivation of the soil, and increase its fertility; for this reason, many farmers grow grains on soil from which roots were harvested the previous season.

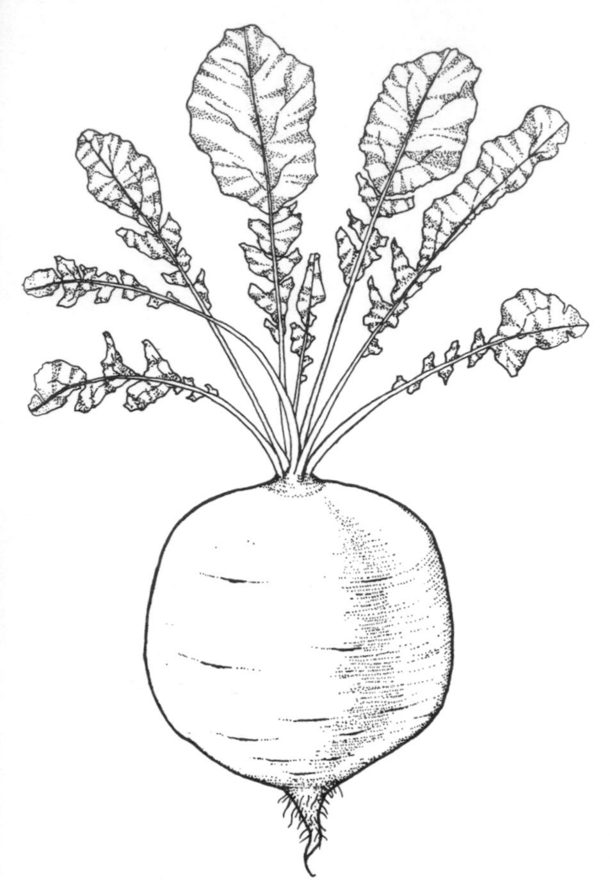

The turnip (above) and swede (below, left) were introduced into England in the 18th century; both are important in crop rotations that include wheat, pasture leys, and barley. The mangold (below, right), another important rotation crop, roots deeper than the others and thrives in drier regions.

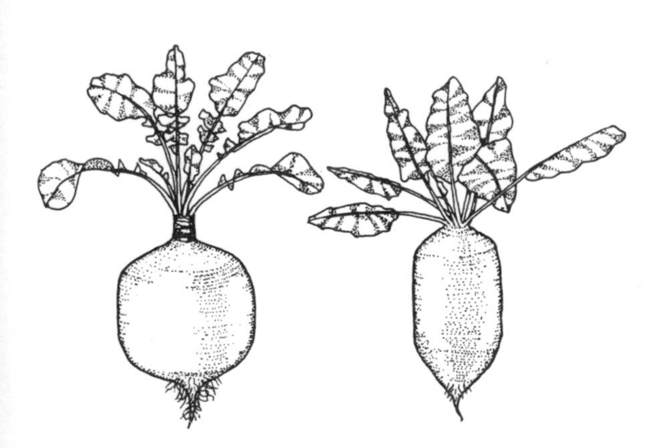

Turnips

The common turnip (*Brassica rapa*), a biennial that probably originated in western Asia, occurs wild in many European countries, especially along the banks of rivers. Its seedling sprouts on bare soil or in mud, and in its first season develops a short stem supporting a cluster of leaves in the shape of a rosette. With the onset of winter it concentrates its food reserves into a large mass of tissues that forms at the point where the primary root joins the stem. Denied food, the leaves and upper stem wither. In the following spring the root sends up a stem bearing yellow flowers, followed by elongated seed pods. The seeds sprout during the summer, and the life cycle begins again. It will be seen that the cycle takes only one year to complete, but since the plant grows during part of two seasons it is classed as a biennial. (This applies mainly to the cold-hardy varieties of the cooler, damper, more northerly countries of the temperate zone. Others, which grow in warmer latitudes such as the southern United States, are true annuals: they are sown in the spring or early summer and, thriving in the favourable climate, develop a marketable root about three inches in diameter within ten weeks.)

The turnip root consists mainly of soft, unlignified xylem cells developed by the cambium. The four main types of turnip are classified according to the shape of the root: *globe* varieties, with a spherical root; *long* varieties, with a root at least three times as long as it is wide; *tankard* varieties, with a somewhat cylindrical shape; and *flat* varieties that are wider than they are long. Turnips are sometimes also classified according to their colour, which varies between white and yellow. They are prepared for the table usually by boiling, their soft, sweet flesh providing a cheap, nutritious food.

A relative of the turnip is the swede (*B. campestris*), which has yellow flesh. Growing in the field, the swede's smooth, bluish-green leaves are easily distinguishable from the rough, green leaves of the turnip; moreover, the upper part of the root of the swede, unlike that of the turnip, is drawn out into a neck that bears the leaves. Both turnip and swede are shallow-rooting plants, and thrive best in damp regions. In drier areas they are often replaced by the mangold (*Beta vulgaris*), a deeper-rooting plant that is better able to withstand drought. The name of this plant is derived from the German *Mangel Wurzel*, "root of scarcity"—a reference to its usefulness when other animal foods are in short supply.

Preparation of the soil for sowing all three species

Three less-common roots. Blackradish (above) is esteemed by some Europeans. Slowbolt (right) is used mainly as a fodder for sheep, its leaves and stem recovering quickly from close-cropping.

Celeriac (below) is a variety of celery; unlike the latter, it is grown for its turnip-like root, which is used in soups.

varies according to the climate. In dry areas the surface of the soil needs to be flat, as for grain crops. In wet areas, however, the surface is prepared as a series of ridges and furrows, the drill planting the seeds on the ridges, which are generally two feet apart. Once the seedlings have sprouted, they are thinned out so as to leave a gap of seven to ten inches between plants.

All three crops give a heavy yield from just a few pounds of seed. Five pounds of turnip seeds, for instance, will yield more than 40 tons of roots. Although only about one tenth of this is dry matter, it still amounts to more than 2000 times the weight of seed used—about double the yield of grains. Mangolds are used solely as animal feed, as is the major share of turnips and swedes. After harvesting, the crops may be eaten on the ground by sheep or pigs, whose dung and urine help to replenish the nutrients that the crops have removed from the soil; or they may be stored in the field in *clamps*—long piles of roots covered with straw and a thick layer of earth—where they will remain in good condition for several months.

Sugar Beet

Along the sea-shores of many European countries there grows a somewhat unattractive, green-flowered weed called sea beet. This perennial sprouts from seed in the spring, grows through the summer, and forms a fleshy root to enable it to survive the winter. The following spring it sends up a flower stalk and the cycle is renewed. Now, an odd feature of sea beet (which belongs to the same species, *Beta vulgaris*, as the mangold) is that the carbohydrate reserves it builds up in its root are not wholly starch (as is the case with most other root crops) but include a pro-

Drawings below show (upper) sea beet, with large, branching roots, and (lower) a typical sugar beet, which was developed from sea beet by selection and cross-breeding. Photograph shows (in actual size) a recently developed variety of sugar beet. Smaller than many varieties in cultivation, it has the unusually high sugar content of 21 per cent.

portion of cane sugar. The first result of man's cultivation and selection of the wild species was the development of the beetroot, a red-fleshed plant with a pleasantly sweet taste.

The rise of sugar beet was triggered off during the Napoleonic war at the beginning of the 19th century, when the British blockaded the coast of Europe and prevented the importation of cane sugar from the West Indies. The idea of extracting sugar from beetroot had originally occurred to a German chemist, A. S. Marggraf, in the 1740s, but he had found that most common varieties of beetroot yield only three per

The sugar-beet harvester below chops off the beet tops, lifts the beets from the soil, and stores them in the container at the back of the vehicle. The crop is loaded onto trucks by means of the side elevator. The farmer's income from the crop is supplemented by the spent beet pulp from the sugar factory and by the tops and leaves, which in some parts of England have replaced turnips and swedes as feed for cattle and sheep.

Sugar beet is unloaded (top) at a factory in Norfolk, one of the major beet-growing counties in Britain. After cleaning, the beets are cut into long slices before being spun in centrifuges (two centre pictures). The centrifuges contain hot water and the extracted sugar emerges as a thick fluid. After refining, the water is evaporated and the resulting sugar crystals are packed in bags (bottom).

cent sugar. With the need to find quick alternatives to cane sugar, however, plant breeders in France and Germany began to select and cross-breed beetroots on a massive scale, concentrating their attention mainly on the more-than-usually sweet White Silesian variety. The result was the sugar beet we know today, which yields about 16 per cent sugar.

Sugar beet thrives on fertile soils in lowlands with warm, sunny summers. The soil should be heavily manured; potash is especially important because it stimulates the plant to increase the proportion of sugar in the carbohydrate. Like mangolds, it favours salty soils—a fact possibly explained by the habitat of its wild ancestor. Being a deep-rooter, sugar beet can tap moisture at a greater depth than grains; its thirst for water is considerable, and each plant needs about 30 gallons during its period of growth. The seeds develop in clusters that cannot be separated without risk of damage. Consequently, the seeds are sown in their clusters, and each is separated from the others when it has sprouted. After the plants have been "singled" they produce a strong head of bright-green foliage, while developing their characteristically swollen root—long, white, and pointed, and quite unlike the beetroot from which it was derived.

The sugar beet, unlike its wild ancestor, is a biennial, developing flowers and seeds in its second year. The farmer harvests his crop during the first year, usually 20 to 30 weeks after sowing. Modern harvesting is carried out by a machine that lifts the plants out of the ground and "tops" their leaves (a nutritious cattle feed) in one operation. But since the farmer has to grow the plant from seed, he leaves a proportion of his crop in the ground until it has flowered and formed seed heads during its second year. Then he selects his seeds from the individual plants whose roots contain the highest concentration of sugar.

After harvesting, the roots are transported to the sugar factory, where they are cleaned and cut by machine into slices. These slices, or *cossettes*, pass to an apparatus in which their sugar is extracted with hot water, which causes the sugar to diffuse through the cell walls. The sugar-water mixture, called "raw juice," is then refined by adding quick lime (CaO), which is afterwards made to settle out of the solution by the application of carbon dioxide. Now the refined syrup is heated and its water content evaporates, leaving behind the familiar white crystals of sugar. About 2 tons of refined sugar can be extracted from 15 tons of sugar beet, which is about the normal yield of one acre.

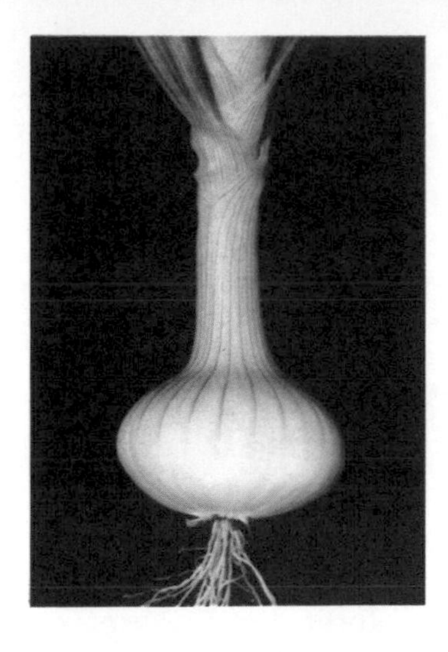

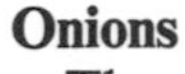

Above (left to right): onion, shallot, and chive, and, below, leek. The storage organs of these and other members of the lily family consist not of roots but of leaf layers enclosing a small bud; their roots extend from the short stem-plate that forms the base of the bulb.

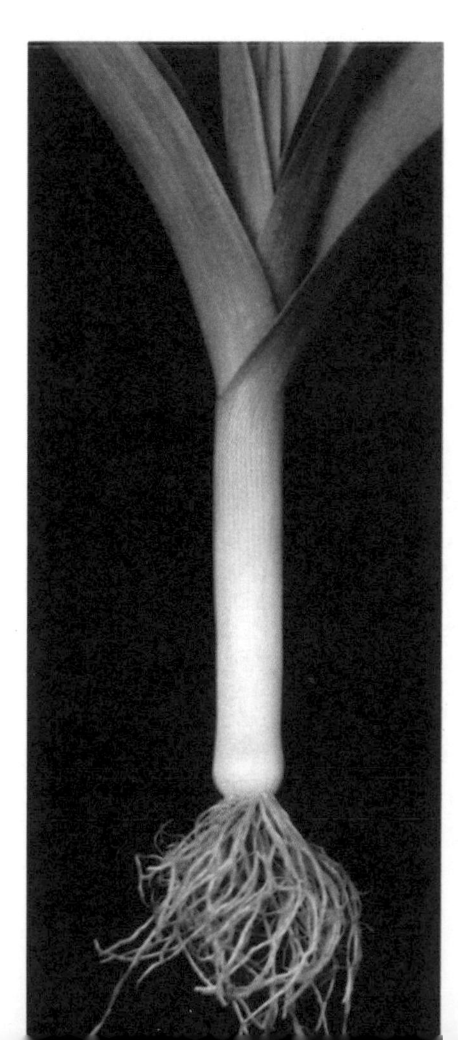

Onions

The onion (*Allium cepa*), another biennial, belongs to the order Liliales, to which lilies and daffodils also belong; related species of the genus *Allium* include the garlic, leek, shallot, and chive. All these species have the curious habit of developing a bulb. Unlike the storage vessels of the plants we have considered so far, the bulb comprises layers of colourless leaves enclosing a small bud. The leaves grow upward from the *stem-plate* at the base of the bulb. They are enclosed within a set of very thin leaves (the onion "skin") that offers some protection against mechanical damage and helps to retain moisture within the plant.

Some bulb plants, such as the shallot, are planted as buds (called *offsets*). Most onions, however, are grown from seed. During its first spring the plant shoots up a tuft of foliage. With the approach of drier summer weather the uppermost leaves wither, and all the foodstuffs manufactured by the plant are concentrated in the bulb, which is then ready for harvesting; as with root crops, a proportion of the crop is left in the ground to provide seeds for the following year. During the second spring, the plant sends up a vigorous seed stalk bearing a compound flower head; lesser stalks, branching out from the central one, also bear flowers—bluish-white in colour and resembling small lilies. When the flowers fade, their papery seed pods ripen, each containing a number of hard, black seeds.

Although they have a definite food value, onions (like other bulb plants) are grown mainly for their pungent flavour, which, paradoxically perhaps, helps to accentuate the flavours of other foods with which they are eaten. Although the onion is a native of western Asia and the Mediterranean region, cultivators in many other parts of the world have managed to adapt the life cycle of the wild plant to regions with much cooler seasonal weather.

Above: young cassava plants, raised from stem cuttings, growing on mounds in a plantation near the Niger River delta.

Below: harvesting the massive cassava tubers. The meal derived from the dried tubers is an important item in the diets of many tropical peoples.

Tropical Root Crops

All the plants we have considered so far in this chapter are either true biennials or are cultivated as such; all are frost-hardy, and therefore suited to temperate climates. In the tropics, however, the main root crops are perennials; their wild ancestors grow in regions that, although never cold, have a marked dry season. The seedlings start life in the rainy season and concentrate their food reserves in tubers or rhizomes in order to survive the dry season. When the rains come again the underground stems send up vigorous flowering shoots that bear seeds but do not entirely exhaust the plants, which form fresh tubers or rhizomes in order to continue the cycle.

Such plants have two main attractions for cultivators. First, their tubers are easy to harvest and store (they do not sprout if they are kept dry), and new crops can be increased by planting tubers instead of seeds. Second, in many regions they can be planted and harvested at various times of the year, so giving a succession of fresh food supplies. Some of these tubers are bitter to the taste, or even poisonous, when eaten raw, but become wholesome after cooking or drying thoroughly. But most of the cultivated varieties have been selected for their sweet taste and absence of harmful properties.

The cassava or tapioca plant (*Manihot esculenta*), one of the most important of this group, is a native of Central and South America, but it is now cultivated in every tropical region, notably in West Africa and Malaysia. Stem cuttings, not tubers, are used for planting and the crop takes about 15 to 18 months to mature. Above ground the fully grown plants form a waving forest of stems, 10 feet tall and topped by clusters of leaves. Below ground they bear huge, fleshy tubers that may yield 12 tons per acre (about 3 tons of dry matter). The cassava root contains hydrocyanic, or prussic, acid, a dangerous poison; but if it is sliced and boiled the poison is rendered harmless and the root can be eaten like a potato. Many of the Indian peoples of South America eliminate the poison by drying the roots in the sun and then pounding them into a coarse meal called *manioc*.

The main substance stored in the cassava root is starch, which is extracted and prepared as the tapioca of commerce. The roots are washed and then fed into a rasping machine consisting of a revolving drum fitted with sawtooth blades. As the roots tumble about in the machine they are scratched to pieces and the starch grains within their cells are released. The starch is washed into settling tanks, and the soft pulp

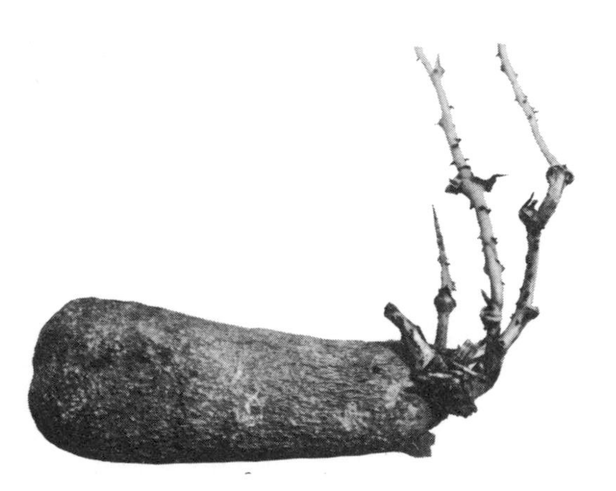

Yams (top, left) are climbing plants and in cultivation are trained to grow up stakes. The roots, seen in storage (top, right) on an eastern Nigerian farm, take about 10 months to reach maturity. The typical root (above) is almost a foot long and as a food is only a little less nourishing than the potato.

that remains behind is dried and used as pig feed. After the starch has settled out of the water it is washed and dried. The round pearls of tapioca that we buy in the shops are formed by rocking the powdery starch in a hammock of cloth. This causes the starch grains to form into round globes, which are then cooked gently to make them hold together.

Other tropical roots include yams (of the genus *Dioscorea*), found mainly in Africa; sweet potato (*Ipomoea batatas*), which is native to Central America; and taro (*Colocasia antiquorum*) of the Pacific islands, from which the thin Polynesian "porridge" called *poi* is made.

The taro corm (above) is an important food on South Pacific islands. The plant, of which several hundred varieties exist, is grown (right) in irrigated soils.

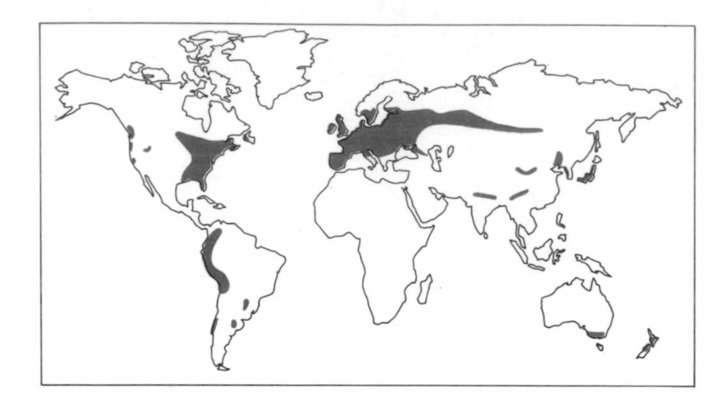

Map (left) shows main potato-growing areas of the world. Indigenous to the moist, equatorial highlands of South America, the potato is now grown mainly in the similarly moist and cool lowland areas of the temperate zone.

Potatoes

The white, or Irish, potato (*Solanum tuberosum*) is a native of the Andean mountains of Peru and Bolivia. Probably the first people to cultivate the plant were Peruvian Indians about the first century A.D., who lived in the High Andes some 12,000 feet above sea level. The first Europeans to encounter the cultivated plant were the Spanish conquistadors who overthrew the Inca empire in the 1530s. It was introduced into Spain some time between 1540 and 1560, and by the end of the century was being cultivated in many parts of Europe. (The word "potato" is a corruption of *batata*, the name given by indigenous peoples of the West Indies to the unrelated sweet potato.)

Although it is native to lands athwart the equator, the potato thrives there only in the cool, damp environment available at high altitudes. This is why it was possible to introduce it so successfully into the similarly cool, moist climate of the northern temperate zone of Europe and North America. The potato's life cycle is governed by cold rather than by shortage of water. It grows on almost any fertile soil except very heavy clay or wet, undrained land. During the warm summer months it sends up vigorous, leafy shoots that develop white, pink, or mauve flowers, followed by green berries resembling small, unripe tomatoes—a plant to which the potato is related. The berries contain small, kidney-shaped seeds that the plant breeder uses for developing new varieties. With the approach of winter the green stems and foliage wither and the plant concentrates its food reserves in its fleshy tubers, which are what we know as potatoes. In peeling potatoes for cooking, the housewife usually removes the "eyes" from the surface of the tuber. These eyes are buds: in the normal growing cycle, the buds break during the spring and give rise to the shoots and tubers of the new season's growth.

For cultivation the farmer uses not seeds but small

Below: 16th-century Spanish sketch of Incan potato harvest. The woman uses a hoe, the man at left a hand plough.

Right: potato cultivation and harvesting in England. After the soil has been well worked, the potatoes are planted, usually between February and May, by a machine (top) that deposits them at the correct interval and covers the seed bed with a mound of earth. The main crop is sprayed in July with a fungicide (centre) to prevent blight. In September the main crop is gathered by a harvester (bottom) that lifts the potatoes out of the ground, separates them from soil and stones, and loads them onto a truck.

tubers called *seed potatoes.* These are planted at intervals of 12 to 15 inches in rows two to three feet apart. After they have grown for a few weeks the stems are "earthed up," to ensure that all the new tubers develop well below the surface of the soil; tubers exposed to light during growth—as often happens with the wild plant—turn green and become unpalatable. The potato is one of the easiest crops to raise, and heavy yields of good quality can be expected as long as the soil has been thoroughly cultivated and dressed with manure or commercial fertilizers before planting. The most important fertilizers are phosphate and potash, which stimulate the development of starch in the tubers. Nitrogen is also important, though very heavy dressings tend to produce potatoes of a watery consistency. The crop is harvested, by hand or machine, when the foliage withers. A good average yield is about eight tons to the acre.

There are several hundred varieties of *S. tuberosum.* They differ considerably in their suitability for different climates and soils, in cooking quality, in resistance to disease, and in appearance. From the commercial standpoint, one of the most important differences is in their rate of growth. According to the time at which they mature, varieties are classed as *first earlies* (early summer), *second earlies* (midsummer), and *main* or *late* crop (autumn and early winter). The first and second earlies are what the housewife knows as "new potatoes." They tend to be tastier and, owing to their shorter growing season, smaller than the main-crop potatoes. In many cases their earliness is due to the grower allowing his seed potatoes to sprout indoors on wooden trays. Given this head start, the potatoes can be planted in April and a good crop harvested in early June. However, they require mild spring weather if they are to provide a heavy first-early harvest. For this reason, most of the early "new potatoes" eaten by northern Europeans are imported from warmer regions farther south, such as the Canary Islands.

The main crop is invariably heavier than the earlies, because the tubers have had longer to develop in the ground: they are planted usually in May and are harvested in October or even later. The main crop not only has to meet immediate needs, but must also be sufficient to satisfy the market throughout the winter and spring.

The chemical composition of a potato tuber depends not only on the variety but also on the conditions under which it is grown. Broadly speaking, water accounts for 75 to 80 per cent of its weight. The remaining dry matter, however, is very nutritious,

Above: this potato riddle on a harvesting machine selects potatoes of the minimum size required by the buyer, very small tubers falling through the mesh. The mesh size can be varied to meet the particular needs of the market.

Below: although clamps are still widely used, many potatoes are nowadays stored indoors, where riddling and bagging can continue in spite of bad weather. This brick store, with forced ventilation, has a capacity of 250 tons.

containing abundant carbohydrate (mainly starch), some protein, and small quantities of many essential minerals. It has been said that an active man could remain healthy on a daily diet of seven pounds of potatoes and a glass of milk. Like many other root crops, potatoes are indigestible if eaten raw. This is because our digestive apparatus cannot break down the cellulose in the cell walls of the tuber; moreover, it can only partly digest the large, hard starch granules. With the application of heat in cooking, however, both the cellulose and the starch swell and soften, enabling our enzymes to break them down into forms that can be digested.

As a world food crop the potato is only a little less important than wheat and rice. World production is about 250 million tons a year, of which the Soviet Union produces about a quarter and Poland one seventh; other important producer countries are West Germany, France, and the United States.

In many European countries in the past, potatoes grown on small plots of land provided the staple food, and sometimes the only food, of the peasantry. If the crops failed, the peasants faced starvation. The classic example of such a disaster was the Irish potato famine of 1846 and 1847, when a million people died and another million and a half were forced to emigrate. The famine was due to late blight, the most widespread and devastating of the many diseases that attack potatoes. Late blight is caused by a fungus, *Phytophthora infestans*. Like other fungi it contains no chlorophyll and so in order to survive it must live as a parasite on plants that are capable of photosynthesis. *P. infestans* selects the potato as host, invading the leaves and destroying them, cell by cell, almost as soon as they emerge above the ground. Deprived of the means to manufacture its food, the potato plant dies. There are possibly 20 or more different races of late blight; many varieties of potato are resistant to one or more races, but few (if any) are resistant to them all. Field research suggests that the endemic home of the late-blight fungus is in the central highlands of Mexico, where all the 16 races so far identified are found in abundance. In a programme of internationally sponsored research, scientists at Toluca, south-west of Mexico City, are attempting to develop entirely new varieties that are resistant to all races. If they succeed they will not only safeguard the diet of peoples all over the world who depend on the potato; they will also make possible its re-introduction into many countries —including Mexico itself—where the depredations of late blight have caused its cultivation to be abandoned.

Above: funeral in an Irish village in 1847 during the famine caused by late blight of the potato crop. Whole towns and villages were abandoned as the people starved to death or, if lucky, managed to emigrate overseas.

Below: potatoes are susceptible to many diseases other than late blight. The picture shows a plant infected with mosaic, a virus disease spread by aphids that causes the leaves to shrivel and so destroys their capacity to make food.

7 Greenstuffs

Man is omnivorous: his diet embraces an enormous number of plant and animal foods; indeed, in the wealthier countries he not only likes but to some extent needs a variety in order to keep healthy, since his appetite becomes jaded if he is confined to a narrow range of foodstuffs. As we have seen, there are many plants that man cannot eat raw but that become tasty and digestible after cooking. There are others, however, that *are* edible in the raw state. They contribute little to the body's total food needs, but they are attractive and beneficial when used as "side dishes" or as garnishing to more substantial foods.

Some of these plants are eaten in very small quantities owing to their strong flavour. One such plant is the spring onion, which is often used in salads. In this early, miniature form (the bulb is rarely more than three quarters of an inch in diameter), the onion is tender and even more pungently flavoured than when fully grown. Another common ingredient of salads is the lettuce (*Lactuca sativa*), the leaves and leaf veins of which are crisp and tender and are easily digested without cooking. *L. sativa* is a native of the Mediterranean region and has been cultivated since the days of the ancient Greeks. It is an annual that grows from seed: given adequate moisture it grows rapidly and is harvested about two months after sowing. It stores food reserves and water in its compact mass of leaves, the inner ones (shaded from the light) being pale and lacking in chlorophyll. The lettuce must be cut when its foliage has reached maximum growth and before the hot, dry, summer weather causes it to *bolt*—that is, to develop flowers. If left too late the plant develops a strong central stem with a flower spike bearing white flowers. From this time on, the stored nourishment in the leaves is absorbed by the flowers for seed-making, and the leaves become tough and indigestible. Commercial growers allow a small fraction of their crop to bolt so as to secure seed for later sowing. The complete life cycle from sowing to harvesting of the seeds takes about three months.

Spring onions (above) are simply immature common onions. Unlike the latter, but like their close relative, the garlic, they are frequently eaten raw, their purpose being to add flavour to other dishes.

Engraving (right), from a botanical work published in 1715, shows different varieties of lettuce, mainly of genus *Lactuca*. The lettuce has been known in Europe since early classical times at least, cultivation of the plant being reported in the 5th century B.C. by the Greek historian Herodotus.

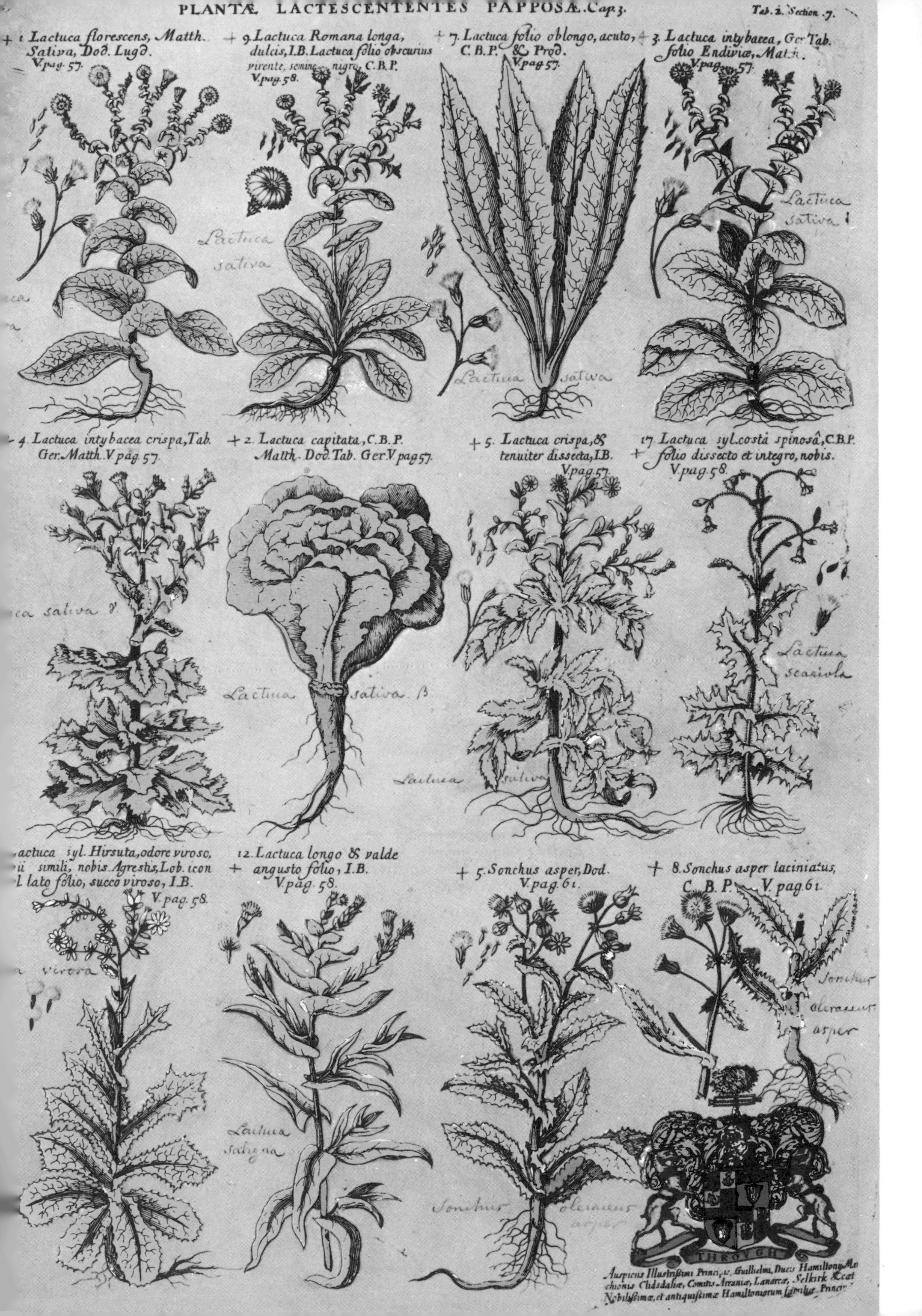
PLANTÆ LACTESCENTENTES PAPPOSÆ. Cap. 3.
Tab. 2. Section. 7.
+ 1 Lactuca florescens, Matth. Sativa, Dod. Lugd. V. pag. 57.
+ 9. Lactuca Romana longa, dulcis, I.B. Lactuca folio obscurius virente, semine nigro, C.B.P. V. pag. 58.
+ 7. Lactuca folio oblongo, acuto, C.B.P. & Prod. V. pag. 57.
+ 3. Lactuca intybacea, Ger. Tab. folio Endiviæ, Matth. V. pag. 57.
Lactuca sativa
Lactuca sativa
Lactuca sativa
Lactuca sativa
4. Lactuca intybacea crispa, Tab. Ger. Matth. V. pag. 57.
+ 2. Lactuca capitata, C.B.P. Matth. Dod. Tab. Ger. V. pag. 57.
+ 5. Lactuca crispa, & tenuiter dissecta, I.B. V. pag. 57.
17. Lactuca syl. costâ spinosâ, C.B.P. + folio dissecto et integro, nobis. V. pag. 58.
Lactuca sativa β
Lactuca sativa
Lactuca scariola
Lactuca syl. Hirsuta, odore viroso, ii simili, nobis. Agrestis, Lob. icon l. lato folio, succo viroso, I.B. V. pag. 58.
12. Lactuca longo & valde + angusto folio, I.B. V. pag. 58.
+ 5. Sonchus asper, Dod. V. pag. 61.
+ 8. Sonchus asper laciniatus, C.B.P. V. pag. 61.
Lactuca saligna
Sonchus oleraceus asper
Sonchus oleraceus asper
THROVGH
Auspiciis Illustrissimi Principis, Guilhelmi, Ducis Hamiltoniæ, Marchionis Clidsdaliæ, Comitis Arraniæ, Lanarcæ, Selkirk &c. Nobilissimæ, et antiquissimæ Hamiltoniorum familiæ, Princip.

Watercress (*Nasturtium officinale*), another common ingredient of salads, has a strong taste reminiscent of mustard (a plant to which it is related) owing to the presence of a sulphur-containing oil in its leaves. One advantage of watercress is that it grows all the year round in temperate climates, even when there is a slight frost. In England watercress beds are commonly found on the edge of chalk downland where perennial springs come to the surface. The plants are set out in beds of mud, which are then flooded. Growth is rapid, until, eventually, each bed is a solid mass of watercress. When the crop is lifted a certain number of offshoots, complete with roots, is held back for the next planting.

We pass now to a somewhat tougher group of plants—the cabbages, which belong to the genus *Brassica* and, like watercress, are related to mustard. Although finely chopped cabbage leaves are often eaten raw, they are more digestible when boiled. The succulence of the cabbage is due to its ability to store water, and the reason for this is interesting. All the many forms of cabbage grown for human and animal consumption have been derived, by selective breeding, from a hardy plant that grows wild in exceptionally dry surroundings. This wild plant is the sea cabbage, which occurs on the coasts of many countries in western Europe, including Britain. It grows in clefts in the bare rock and is often immersed in salt spray thrown up by breaking waves. The presence of salt makes it very difficult for a plant to draw water from the soil, while the drying effect of the sun's rays, intensified by reflection from the sea, adds to the plant's problem of conserving water. In short, the sea cabbage is obliged to live rather like a desert plant, such as a cactus, storing any moisture it collects during rainy spells within its swollen, succulent stems and leaves. To aid storage the cuticle, or surface-cell layer, of its leaves is very waxy and quite waterproof; this prevents water from being sucked osmotically out of the leaves by salt from the sea spray. The characteristics of the cabbage tribe are thus determined by their original habitat; but man has taken the plant and grown it under quite different conditions—away from sea salt and in fertile soil. The succulence and resistance to drought remain in the cultivated varieties, but the crop yield is immensely bigger.

All the cabbages are biennials, making their leaf growth during the first year and flowering and seeding in the second. Apart from maintaining a supply of seed, our interest is almost entirely in the first-year growth. Part of its importance lies in the fact that the

Harvesting watercress (above) in early March on a farm in Lincolnshire. The plants are raised here in large concrete tanks fed with water from artesian springs nearby.

Engraving (right) is from *Paradisi in Sole*, a book on flower and vegetable gardens published in 1629, and shows eight members of the cabbage tribe: (1) closed and (2) open cabbage, (3) savoy, (4) cauliflower, (5, 6, and 7) various coleworts, (8) kohlrabi.

1
2
3
4
5
6
7
8

Above, left: cabbage seedlings, having been raised in a nursery plot, are planted by machine. When harvesting, the farmer leaves a proportion of his crop in the ground so that they will provide seeds for the next season. The seeding cabbage (above) is wrapped in a linen bag to protect the seeds against birds.

crop is available in the autumn and winter, when other vegetables are scarce. The typical cabbage for human consumption grows a hard, spherical mass of leaves arising from an extremely short stem. The outer leaves, exposed to the light, contain more chlorophyll than the inner ones, which are pale yellow or almost white.

Most members of the cabbage tribe can be sown by drilling the seed into a well-manured field. It is sometimes more convenient, however, to sow the seeds first in a nursery bed and to transplant them a few weeks later. In many regions, planting out is done by hand. But during the last 25 years or so machines have been developed that plant the young cabbages and firm up the soil around their roots. However, the larger firms of market gardeners that use mechanical planters still cut the crop by hand. Most growers plant several different varieties of cabbage; the quicker-growing varieties are ready by autumn and the slower-growing ones, which are frost-hardy, are cut during the winter. A yield of 40 to 50 tons to the acre can be expected.

Even the hardiest cabbages cannot withstand the severe winters that commonly occur in many parts of north-western Europe; and this poses an acute problem of storage. One can store cabbages in sand, as is done with carrots; but it is a clumsy method and extremely unpopular with the housewife, who has to clean the vegetable before cooking. In many countries —notably Germany and Scandinavia—an alternative method is used. The product of this method is best

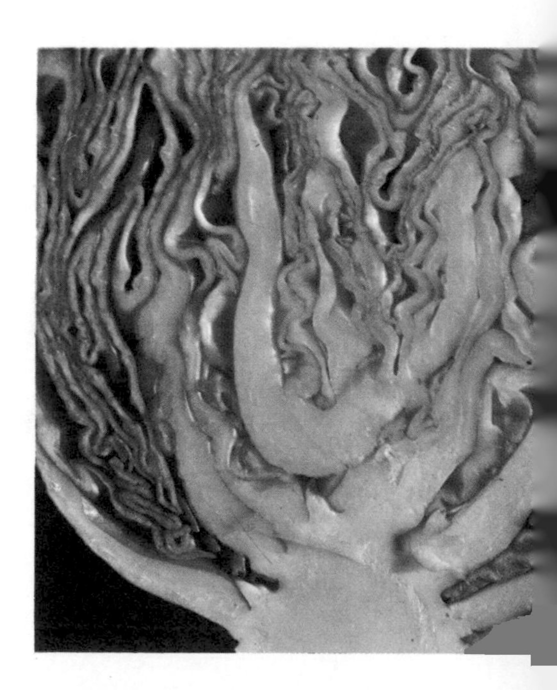

The leaves of some sea cabbages are tinted red. Selection and cultivation of individual plants with this feature resulted in the red cabbage (above), which is used mainly for pickling.

Above: Brussels sprouts, with a cluster of miniature cabbage heads beneath a crown of photosynthetic leaves.

Kohlrabi (below) is valued not for its leaves but for its fleshy, spherical lower stem. Kale (below, right) is a primitive form of cabbage that has been little modified by cultivation. It is used mainly as a winter feed for cattle.

known by its German name, *Sauerkraut* ("sour cabbage"). Newly cut cabbages are shredded and packed into barrels with alternating layers of rock salt; the layers are rammed down firmly, to exclude air. The cabbage now undergoes a fermentation process that lasts for several weeks, after which it can be eaten raw or boiled.

One rather curious form of cabbage, the Brussels sprout, develops not a solid head of leaves but a number of miniature heads from a long central stem. These sprouts are just like miniature cabbages; but, unlike the large cabbage, their flavour seems to be improved by exposure to frost. In Britain, toward the end of a hard winter, they are more plentiful than any other cabbage and their slightly nutty flavour can become monotonous. (The story goes that, during World War II, when alternative vegetables were very hard to come by, an American bomber crew stationed in Norfolk was ordered to destroy several thousand acres of sprouts in Lincolnshire in order to relieve the jaded palates of their colleagues.)

Several members of the cabbage tribe make excellent winter feed for stock. One of these is kohlrabi, an odd-looking plant that, like root crops, stores water and foodstuffs in a swollen stem; it can be grazed in the field or lifted and stored in clamps like turnips. Then there are the kales, tall-growing cabbages that flourish throughout a mild winter. Kale is often grazed directly, the farmer using an electric fence to confine the stock to their rations for the day and to

Pictures show the dense cluster of the cauliflower inflorescence from above and in cross-section. Though indigenous to the Middle East, the cultivated plant thrives best in a cool, moist climate.

prevent them from trampling the rest of the crop. But, like silage and hay, it is also an important source of feed for cattle wintering under cover.

All the greenstuffs we have considered so far are harvested or grazed when their foodstuffs are concentrated in the stems and leaves. But there are a few important exceptions to this, including cauliflower and broccoli (the latter named after the man who pioneered their cultivation). These cabbages develop a close mass of flower buds within a protective casing of leaves. Cauliflowers make a welcome change from ordinary cabbage, especially when garnished with sauces.

With all these greenstuffs, as with the other crops we have looked at so far in this book, the objective of the grower is to provide the best possible *natural* conditions for the plants in order that they may produce the highest possible yields; the role of the grower, in other words, is not so much to change nature as to improve upon it. But there are some plants, necessarily grown in small amounts, that are good to eat (and expensive to buy) provided they are grown under *unnatural* conditions. The basic idea is to deprive the plants of daylight (and therefore the opportunity for photosynthesis), so forcing them to rely for food and growth on the materials they take in *via* their roots. The reason for this seemingly odd procedure is that these plants are tasty and tender if they are prevented from manufacturing chlorophyll, but that as soon as photosynthesis begins they become tough and bitter. One such plant is asparagus (*Asparagus officinalis*), a member of the lily family. Originally a wild, sea-coast perennial, it is now grown in well-manured trenches. As soon as a new shoot appears it is covered with soil, so that the plant is continuously pushing upward in an attempt to reach the light. The result is a tasty stem stick, of which only the growing point at the top is slightly green-tinged; the rest, lacking chlorophyll, is creamy white, or *etiolated*, as the biologist calls it.

Another plant that is interesting to eat only when etiolated is rhubarb (*Rheum rhaponticum*). In its wild state rhubarb is an Asiatic perennial with a large woody root, red stems, and very large green leaves. The stems are tough and acidic, and not worth eating even when cooked, while the leaves are so bitter that even animals reject them. However, the cultivated rhubarb we buy in the shops is grown by planting root stocks in dark-houses without windows, and by applying gentle heat in springtime to encourage their growth. The result is the familiar pale-pink stems, topped by etiolated leaves. When stewed, the stem (but not the leaves) has a delicate, fruity taste.

Cutting asparagus sticks (above, left) on a Cambridgeshire farm. The use of the saw-toothed knife prevents the cut stem from "bleeding." The ripened sticks (above, right) are creamy white because they have been forced to grow beneath the surface of the soil.

To eliminate the normally fibrous texture and acidic taste of the stems, most commercially grown rhubarb is raised in dark-houses (right). As in the case of asparagus, the absence of light leads to the development of tender, tasty stems.

8 Soft-stemmed Seed and Fruit Crops

There are three main groups of plants that yield substantial crops of seeds or fruits that are good to eat: the tough-stemmed grain grasses (Chapter 3), the woody-stemmed orchard trees (Chapter 9), and a large group of plants with soft stems. Many of these soft-stemmed plants are annuals, or at least are treated as such in cultivation: they grow from a single seed to a mature plant bearing ripe seed within the space of a few months. Others have biennial stems that bear fruit in their second season, and then wither; the root systems that bear them, however, are perennial.

All the soft-stemmed fruit and seed crops demand careful tending by the grower. They need fertile, thoroughly weeded soil and a good deal of sunshine for their rapid growth. Many are climbing plants that must be supported by branchwood, poles, or wires.

Scarlet runner bean (above) has the five-petalled flowers characteristic of the Papilionaceae. Like most large-seeded species of the genus *Phaseolus* it is a native of the Americas.

Peas and Beans

All the many different species of peas and beans belong to the natural family Leguminosae, which is one of the largest in the plant kingdom and includes several hundred seed and fodder plants useful to man. In general terms, those legumes with large, spherical seeds are called peas, and those with large, flat or oblong seeds are called beans. Some of these plants are native to America, some to Europe, and some to Asia; but most of the more valuable species have been introduced into every country where they can thrive.

The subfamily of the Leguminosae to which peas and beans belong is called the Papilionaceae (from Latin words meaning "butterfly flowered," because their flowers resemble butterflies resting on leaves and stalks). Each flower has five petals: one large upright one at the back called the "standard"; two medium-sized flat ones spreading out to each side, called the "wings"; and two small bent ones, folded together at the front, called the "keel" (they form a shape rather like the keel of a boat).

Photograph (right) shows 16 of the many hundred different kinds of pulses (seeds of leguminous plants) cultivated by man.

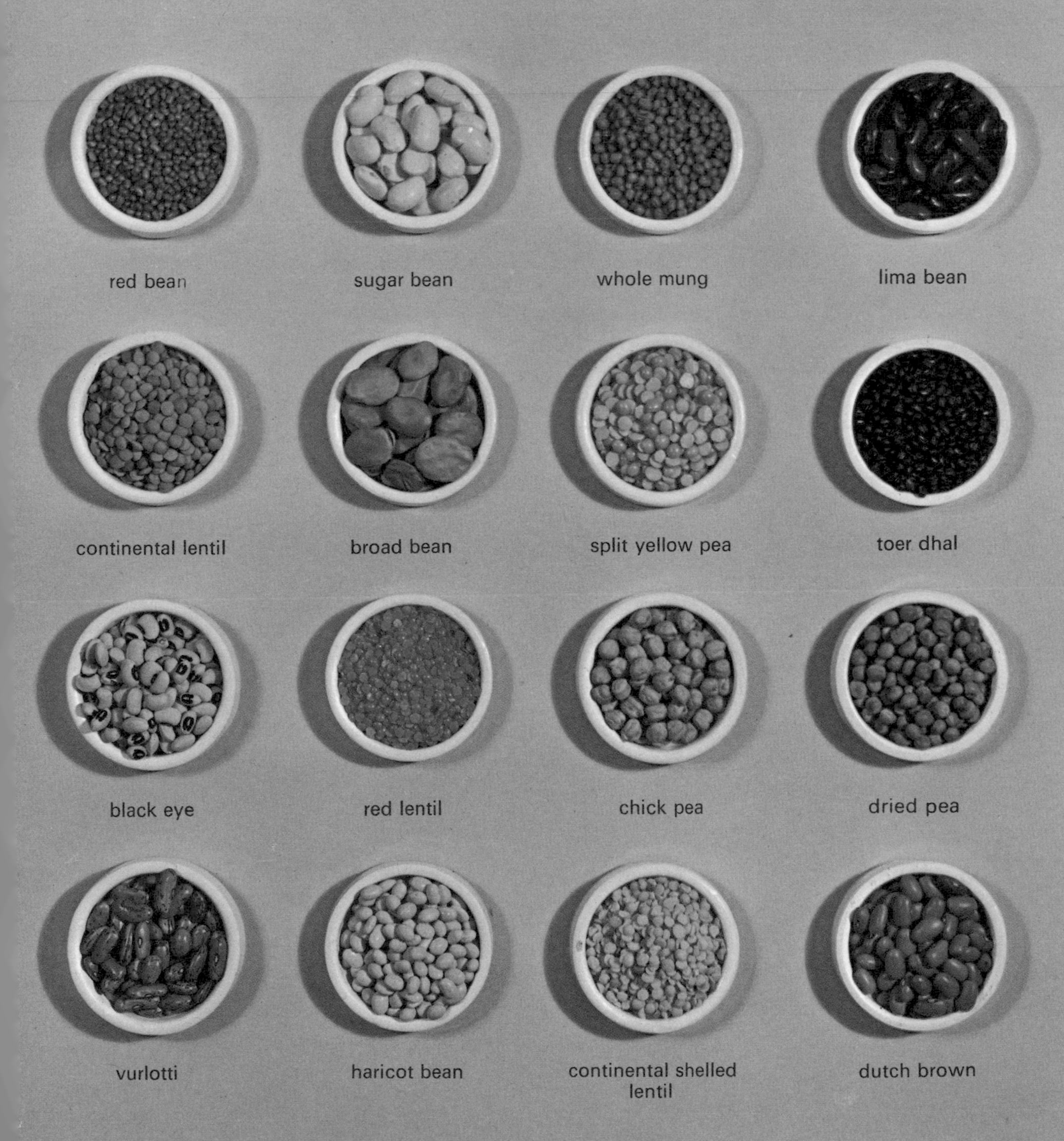
red bean
sugar bean
whole mung
lima bean
continental lentil
broad bean
split yellow pea
toer dhal
black eye
red lentil
chick pea
dried pea
vurlotti
haricot bean
continental shelled lentil
dutch brown

The green pea (above) is cultivated in the temperate zones of North America, Europe, and Asia; by far the largest tonnage is grown in northern China.

Below: night work on a specialist pea farm, where viners remove the peas from their pods. Peas (inset) emerge from the freezer at −20°F before being packed.

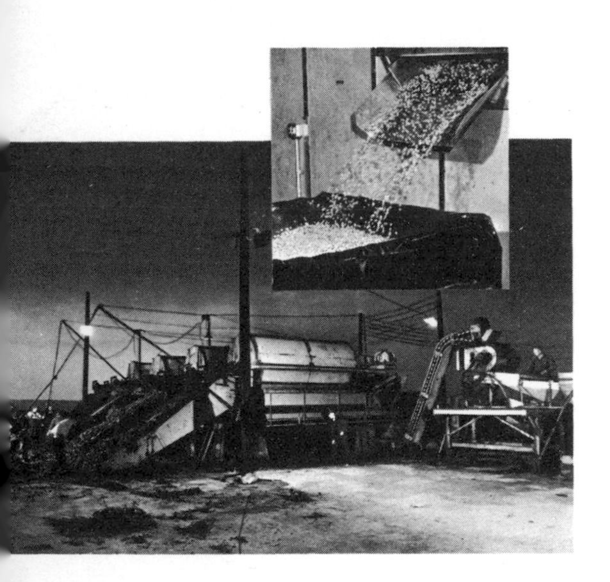

We saw in Chapter 1 that legumes are an important source of protein owing to the presence of nitrogen-fixing bacteria that live in nodules in the plants' roots. In some of the poorer countries peas and beans are the main source of protein in the human diet because it is easier and cheaper to grow them than to keep large numbers of animals. Few peoples rely entirely on plants for their proteins, however; most add some kind of animal protein to their diet in the form of milk, eggs, fish, or meat.

One of the most widely grown leguminous plants is the green pea (*Pisum sativum*), so-called because it is usually eaten in its "green" (that is, unripe) state, while it is still soft and sweet. Several varieties are grown, harvested, and prepared for the table in different ways. All are descended from wild ancestors that probably originated in central Asia and have been cultivated since early classical times. In market gardens the green pea is grown as a vine, or climbing plant. Branchwood is stuck into the ground to support it, and the plant scrambles up into the sunlight by means of slender tendrils (modified leaves) that grasp the branchwood and may grow to a height of 12 feet or more. The seeds form in long pods that are harvested while still green. At this stage the seeds are rich in protein and sugar and so are nutritious as well as tasty.

Green peas are also grown as field crops; usually the grower plants varieties that form bushes and so do not need to be supported. At one time field-grown peas were left to ripen fully because they could be stored and sent to the market only in the hard state. Nowadays, however, they are harvested by machines, called *viners*, that can free the soft, unripe peas from their pods without damaging them; then they are canned or quick-frozen, so that they are available to the housewife throughout the year.

Commercially, the most important beans belong to the genus *Phaseolus*. The runner bean (*P. multiflorus*) is so-called because its stalk "runs" or climbs, unlike the many species of "dwarf" bean. The runner does not have tendrils like the green pea but climbs upward by winding its stalk round and round a pole (or, in the wild, around the stem of another plant). In warm summer weather the young bean-stalks shoot up the poles in a few weeks and bear bright-scarlet flowers. After pollination, long green pods develop containing the pink seeds, or beans. Before either pods or seeds have hardened, they are harvested for eating in their green state. At this time the food materials ultimately destined for the seeds are concentrated mainly in the pods; indeed, it is really the pods that form the bulk of the runner-bean harvest, because the seeds are still too small to count for much. Until the quick-freezing process was developed runner beans were commercially available only during a short summer season; today they are available, like green peas, all the year round.

Many species of beans are grown in the field and allowed to ripen fully. Their hard seeds can then be threshed out of their pods like grain and stored indefinitely in a dry atmosphere. Although in many countries the field crops are still cultivated by hand, in North America and in some parts of Europe sowing, weeding, and harvesting have been fully mechanized. The haricot bean (*P. vulgaris*), which has been cultivated in North and South America for at least 1000 years, is one of the most important of these field crops. In its baked and canned form it is exported all over the world and eaten as a cheap, pleasantly flavoured alternative to meat.

The soybean (*Glycine soja*) has been cultivated in its native China for more than 4000 years, and is the most important source of plant protein in both China and Japan. Today, however, about half the world's tonnage is produced by the United States, for the value of the soybean now extends far beyond its original use as a vegetable. Its commercial importance lies in its high concentration of edible oils, which are used in the manufacture of margarine and cooking fats and which are also important raw materials in the production of paints, enamels, cosmetics, and adhesives. The meal remaining after the oils have been extracted provides a high-protein feed for livestock.

The peanut or groundnut (*Arachis hypogaea*) is quite distinct from the "true" nuts borne as single-seeded fruits on woody-stemmed trees. It is simply the hard, ripe seed of a leguminous plant—one of many seeds that ripen within a pale-brown, papery pod. A remarkable feature of the peanut plant is its habit of burying its own seeds. It is a low, bushy plant bearing the papilionaceous flowers typical of the subfamily. As its seeds ripen the stems bend downward and force the pods into the ground; in this way the seeds are not only protected but sown for the next season's growth. Originally a native of South America (where it has been cultivated since pre-Columbian times), the peanut is now grown extensively in North America, in Africa, and in many other countries with subtropical climates. The seeds are eaten raw or after roasting, or are made into peanut butter; a high proportion of the annual harvest, however, is grown for the oil in the seeds, which is extracted by crushing and used in cookery.

Stems and leaves of the peanut (above) are often cut for hay, while in its native South America the unripe nuts and pods are used as hog feed.

Africa is one of the largest producers of peanuts. Most of the massive harvest (below) at Kano, northern Nigeria, is exported to the temperate zone.

In many commercial nurseries tomatoes are raised entirely under glass. The plants are scramblers and must be supported by stakes or trellises. When the tomatoes begin to develop the lower leaves are cut off to concentrate growth in the fruit.

The "automatic bee" (above), a device for collecting pollen, is shown at Wye Agricultural College, where it is being used to develop new strains of tomatoes. Pollen from one variety fertilizes another; fruit from the resulting hybrid are seen (above, right). If the hybrid has commercial potential, the next stage is to grow a seed crop.

Fruit from the new variety are crushed and the seeds separated from the rest of the pulp. About five to seven ounces of seeds can be got from 100 pounds of fruit.

Annual Soft-Stemmed Fruits

Many soft, pleasantly flavoured fruits are borne by annual plants that can sprout, flower, and bear seed only with the help of a hot, sunny summer. The wild ancestors of all such soft-stemmed fruits are native to the tropics or subtropics; none of them can withstand frost and so they cannot grow wild in the temperate zone. Cultivated varieties are raised in temperate climates, however, as *half-hardy annuals*—which means they are grown only during the frost-free summer months and are "preserved" through the winter in the form of seed for the next year's sowing.

The tomato (*Lycopersicum esculentum*), the best-known fruit in this group, is a native of South and Central America, and takes its English name from the Amerindian word *tomatl.* It had been cultivated for several hundred years when it was "discovered" by Spanish conquistadors in the early 16th century. Europeans did not, at first, care for its distinctive flavour: when it was introduced into Europe it was grown mainly as an ornament, because many people thought it was poisonous. The first Europeans to exploit its culinary value were the Italians, who began to use it to flavour sauces. From the 18th century onward, its use spread to other countries, both in Europe and in America, until, by the end of the 19th century, it had become the mainstay of nursery and greenhouse industries in many parts of the world.

In the tropics tomatoes are grown out of doors all the year round. In cooler regions the seed is sown in

Glass-fibre troughs (above) are used for raising tomatoes at the Glasshouse Crops Research Institute. The plants grow in a loam-compost medium, nutrients and water being fed in through perforated pipes on the bottom of each trough.

The cucumber plant (below) has long, trailing stems that grow horizontally in the wild. In cultivation, as here, they are trained up trellises.

greenhouses during the spring; later the young plants are either transplanted in the field or continue to maturity under glass. The tomato is really a vine or scrambling plant and has to be supported by stakes. It needs a rich soil, ample water, and careful tending. As the fruits ripen from green to red, the leaves are removed so that sunlight can reach every part of the plant. Each fruit contains many flat, pale-brown seeds, which are able to pass undamaged through the digestive systems of many animals, including man, and so in some countries tomato eaters unwittingly help to propagate the plant.

The tomato is one of many species that lend themselves to *hydroponics*—a method of cultivation in which plants are raised in a solution that contains all the nutrients essential to their growth. One advantage of this method, which is carried out under glass, is that the grower strictly controls the foodstuffs taken in by the plant and thus is able, in some measure, to determine its cycle of growth. A somewhat similar method of tomato cultivation has recently been studied by the Glasshouse Crops Research Institute at Littlehampton, Sussex. The seedlings—which are planted at a density about six times as high as with conventional methods —are set in earth or a synthetic medium placed in long troughs made of glass-fibre; the seedlings are supplied with water and nutrients from a perforated tube running along the bottom of the trough. The plants are allowed to grow until they have developed their first truss, or bunch, of fruit; thereafter, all further upward growth, and any side shoots, are suppressed, so that the plant's food is concentrated in the single truss. It has been found that this method raises the crop yield from the present average of about 40 tons to as much as 120 tons per acre.

The tomato, besides having a rich flavour, contains sugars and vitamins—it is a useful source of ascorbic acid (Vitamin C). The raw fruit is widely used in salads; it is also canned and bottled as sauce, juice, and purée on a very large scale.

Cucumbers, marrows, squashes, pumpkins, and gourds all belong to a large natural order called the Cucurbitaceae (from the Latin *cucurbita*, meaning a gourd). Some are native to Europe, Asia, and Africa; others to America. But all are now cultivated in every country of the Old and New Worlds in which it is warm enough for them to thrive. Cucurbitaceae have long, trailing stems that give the wild-growing species an advantage over other plants: they are able to scramble over their competitors and so gain the benefit of every hour of sunlight. As a consequence,

they develop and ripen large, succulent fruits very quickly—often after only a few months of growth. The cucumber itself (*Cucumis sativus*), which is a native of India, is usually harvested between 60 and 70 days after planting. In the temperate zone the Cucurbitaceae are usually grown as climbers in greenhouses; sometimes "hotbeds"—heaps of fermenting manure—are used to give the plants the extra warmth they need for rapid growth. Their fruits are largely water, but they also contain some nourishing sugars. Some of the large African gourds have tough, woody skins that are dried and used as food- or water-containers after their soft pulp has been removed and eaten. Many Cucurbitaceae fruits are eaten raw; others are cooked, often with meat stuffing; still others, such as gherkins (which include the immature common cucumber fruit), are pickled in vinegar.

Harvesting gourds (above) in Cameroon. Below: the tough outer skins of gourds have been adapted to make a Cypriot rattle (left), an Indian sitar (centre), and a South African bow harp (right).

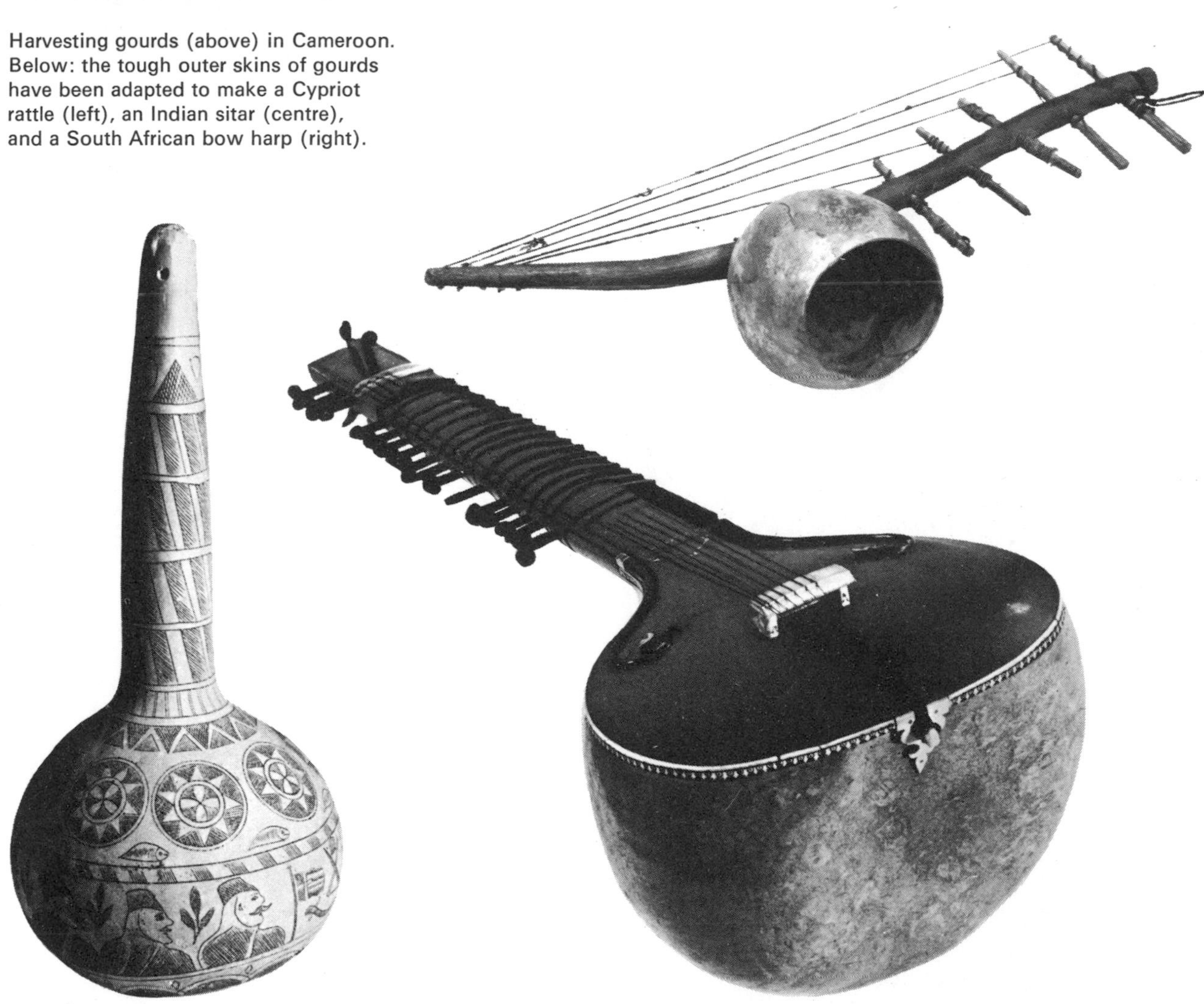

Biennial Soft-Stemmed Fruits

Many of the species in this group lead, as it were, a double life. Each summer they develop tall stems that bear flowers and fruit every second year and then wither. Below ground, however, they do not behave as biennials: at the beginning of their third year they develop a fresh growth of biennial stems that spring up from *perennial* underground stems or roots.

The raspberry (which includes several species of the genus *Rubus*—a member of the rose family) is a good example of this growing habit. In its wild state it is a low, shrubby plant commonly found in the woodlands of North America and northern Europe. Under cultivation its slender, upright shoots are trained up a framework of wires. The shoots spring up freely every spring, either as *suckers* from underground roots or as offshoots from the base of an old stem. In their second year they bear white flowers, followed by a heavy crop of red berries, each consisting of a cluster of pulpy seeds, or *drupelets*. The berries are harvested for eating raw (either immediately or after freezing), for preserving whole in sugar syrup, or for making into jam.

Wild blackberries, which are closely related to the raspberry, grow in a similar way but form huge, straggling bushes; even the many cultivated varieties

Above: blackberry (left) and strawberry. The blackberry's seeds are in the pulpy drupelets. The strawberry has *achenes*, or fruits, embedded in its surface.

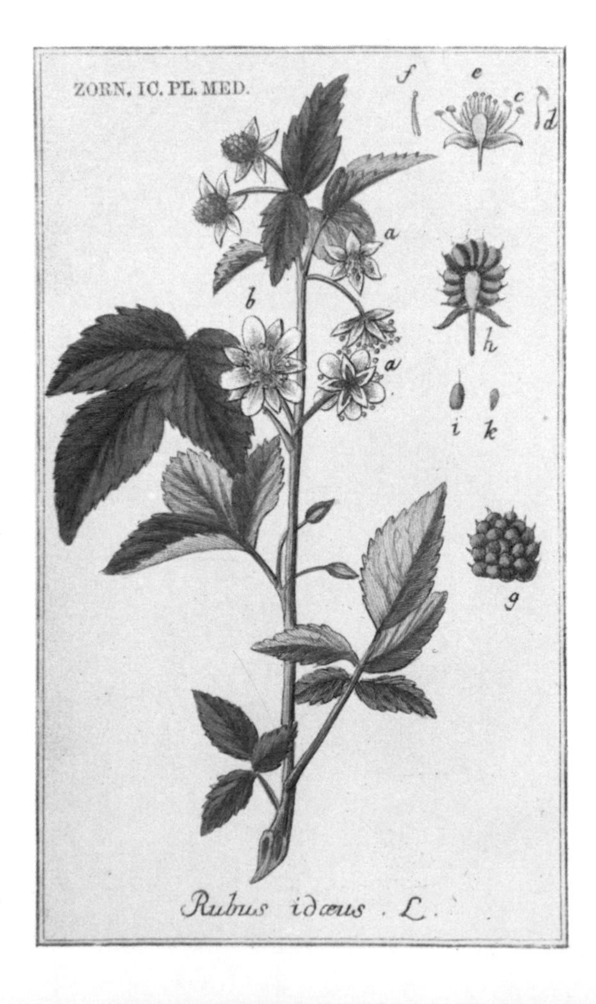

Below: blackberry (left) and raspberry, both members of the bramble tribe.

Loganberry (above) and wild strawberry (below). The strawberry plant spreads by means of its tentacle-like stems, or runners, which touch the soil and take root.

must be drastically pruned to keep them in check. One of the reasons for this is their habit of natural *layering*: any branch that touches the ground can take root, and this enables the bramble thickets to spread very rapidly. The loganberry, which has dark, wine-red fruits, is a cross between a raspberry and a blackberry; it was originally bred from seed in California in 1881 by Judge J. H. Logan.

Wild strawberries (members of the genus *Fragaria*, which also belongs to the rose family) are common on dry, wooded downlands in northern Europe, and bear white flowers followed by red fruits on short, tufted plants. The strawberry is a true perennial and spreads by means of runners (slender stems) that take root when they touch the soil a foot or so away from the parent plant. The strawberry owes its name to the custom of cultivators of laying straw around the base of each plant in order to keep the low-growing fruits clear of the soil. Most cultivated varieties are the result of crossing wild European and wild South American species. Today, they probably have a wider distribution than any other commercially valuable perennial fruit plant: they are grown from the subarctic to the cooler tropical regions, suitable varieties having been developed for each climatic zone.

Banana

Although the banana plant looks very much like a tree, it is actually a giant herbaceous plant with a rather curious method of growth. Below ground it develops a rhizome that bears roots and lasts for many years. The rhizome develops annually an upright, hollow "false stem" that resembles the trunk of a tree but in fact consists of a rising succession of leaf stalks, each folded within the next beneath it. The false stem is crowned with up to 20 quill-shaped leaves that may attain a length of 12 feet and are among the largest in the plant kingdom.

After this great rosette of leaves has developed, there grows up within the false stem, hidden from sight, a flower spike, or "true stem," that bears flowers and fruit. It emerges through the top of the false stem and then turns downward. At this point it develops radial clusters of female flowers and, farther down, clusters of male flowers; it ends in a bud. After pollination, the female flowers develop into the familiar banana fruits, which form clusters, called "hands," around the stout spike. Once the fruits have ripened, the entire structure of leaves and false stem withers and dies back to the ground—although in cultivation it is usually cut down before this happens. It is replaced by other leafy shoots springing up from the rhizome, so that an individual plant will continue to bear fruit for several seasons. New plants are raised by cutting off some of these shoots and transplanting them.

Wild banana plants grow in the jungles of tropical Asia in areas where enough sunshine can penetrate the dense canopy of taller timber trees. These wild plants bear small, straight fruit that are much less sweet and tasty than the cultivated varieties, though they are a useful source of food to the tribesmen (and monkeys) that still inhabit the remoter jungle regions. The wild fruits develop small seeds, whereas most cultivated species (*Musa sapientum* is the most important) are seedless.

The peoples of India, Indonesia, and Malaysia brought the banana into cultivation several thousand years ago as an orchard crop to be grown around their villages, and by selection gradually developed varieties with large, well-flavoured fruits. Today the typical Malaysian fruit market boasts a remarkable range of bananas of different shapes, colours, and flavours. Probably during the first few hundred years of the Christian era, Arab traders introduced the banana into Africa (its name is African in origin); after the discovery of the Americas it was imported into the Caribbean region—now the main source of the world

The banana plant blossoms nine or ten months after planting. The flower spike (above) is about 18 inches long at this stage; within its green sheath-like bracts are circular clusters of female flowers that later will bear bananas.

Below: the bracts sheathing the flower spike are cut away to expose the young bananas. Each cluster, or hand, on the spike carries 10 to 18 bananas, and each spike develops between 6 and 12 hands. The crop is harvested, while still green, 12 to 15 months after planting.

The ripening bananas (right) bend upward to get full benefit of the sunlight. The bare stalk below once bore male flowers. After cutting, the crop is dipped (above) in various solutions to neutralize and remove every trace of insecticides.

Bananas deteriorate rapidly when ripe, so they are picked while still green and are ripened (below) under controlled conditions in countries in the temperate zone to which they are exported.

market—from the Canary Islands. When the peoples of Europe and North America developed a liking for bananas it was at first possible to meet their needs only from the nearest tropical lands—mainly the Canaries for Europe and Cuba for North America. The bananas were cut while they were still green and ripened during the course of the short sea voyage. Nowadays, transport and storage are more highly developed; huge cargoes of unripe bananas are shipped at carefully controlled temperatures from Jamaica to Britain and from Costa Rica to the United States, and after their arrival are ripened by artificial heat. Banana-lovers in the temperate zone (the main export market) prefer yellow, curved fruits, so varieties bearing this type of fruit predominate in the largest tropical plantations. Although the banana holds much water it is a very nutritious fruit, most varieties containing up to 22 per cent carbohydrate (mainly fruit sugars) and several vitamins and minerals.

Pineapple plant (above, left) with long, spiked leaves and immature fruit. The ripe pineapple (above) consists of a mass of fruits, fruit stalks, and the central stem from which they grow. It is, however, usually harvested before the fruits have separated, as here.

Pineapple

A short-stemmed herbaceous plant, the pineapple is well adapted to living on the arid grasslands of its native Paraguay and southern Brazil. Its 30 or 40 slender leaves, two to four feet long, are tough and bluish-green and are able to retain moisture and to reflect the fierce rays of the sun; little spikes along the leaf edges protect them from grazing animals. The pineapple that develops (about 18 months after sowing) in the heart of this tuft of leaves is not a simple fruit but a swollen stem comprising a mass of many small fruits. It ends in a crown of short leaves that, if cut and planted, will take root and form a new plant.

The pineapple (*Ananas sativus* or *comosus*) was first cultivated by South American Indians in and around its countries of origin. Its area of cultivation spread northward until it was introduced into the Caribbean islands by migrating Arawak or Carib Indians. Columbus received pineapples in barter trade from the natives of Guadeloupe Island in 1493, and the fruit

Machine (above) lays three rows at a time of mulch paper on a pineapple plantation. The paper, which resembles tar roofing felt, discourages weeds and holds moisture and heat near the soil surface. The marks on the paper indicate the position of the pineapple "slips," which are hand planted (above, right).

At many factories, pineapples are cut by machine into cylinders and then pass by conveyor to trimmers (below) who remove any remaining bits of skin. The fruits are then sliced and canned.

rapidly became popular with Europeans. During the 16th century the Portuguese introduced it into India and Indonesia. (It is quite possible that the introduction was accidental: the fruit was often carried by the Portuguese as a refreshing shipboard food, and it may be that the first Indian and Indonesian stocks arose from crown leaves tossed away by voyagers to these lands.) In Europe, French and English gardeners grew it in green-houses and in time developed exceptionally good varieties that were used as the foundation stocks for the Hawaiian and Australian pineapple industries established toward the end of the 19th century.

The pineapple is now grown in most tropical and subtropical countries, notably the Hawaiian Islands (where half the world's export crop is raised), Indonesia, Malaysia, South Africa, and northern Australia. It is easily propagated by transplanting the offshoots that develop at the base of each main stem (the crown leaves are never used in cultivation). The pineapple thrives best in a dry atmosphere and in rather poor, light soil enriched with fertilizers. In the Hawaiian Islands and elsewhere, long strips of bitumen-impregnated paper are laid by machine along the rows in which the pineapple offshoots, or seed pieces, are to be planted. The paper acts as a *mulch*: it helps to reduce evaporation of moisture from the soil surface and also discourages the growth of weeds. When a fruit is harvested, the stem that is left branches and later bears two fruits; but these are smaller than the original fruit, so the process cannot be repeated indefinitely by the commercial grower. Many pineapples are eaten in their countries of origin; some are exported whole, mainly to countries in the temperate zone; but most of the world crop is canned, as slices in sugar syrup or as juice, so that it can be stored.

9 Orchards and Vineyards

The soft fruits and hard nuts borne by woody-stemmed trees and bushes are among the oldest foods known to man. Our hunter-gatherer ancestors, living at the edges of the great tropical forests, knew how to distinguish between tasty and inedible species, and where and when the tasty ones could be found. Their lives depended on such knowledge because they competed for these foods with many species of animals, including birds, monkeys, and squirrels, as well as with their fellow men. Often, in times of shortage, groups of people would fight for possession of the meagre harvest. The ability of one group to defend a grove against other, less-powerful, groups led eventually to the idea of communal ownership. In some tropical lands, groups or tribes of nomadic food-gatherers still regard groves of wild fruit- and seed-bearing trees as their communal property, and will fight any other group that tries to take a share of the annual harvest. Sometimes a grove of such trees is the only fixed holding of land that a tribe claims to own. The people often move their huts or shelters from place to place but never wander far from their traditional source of nuts and fruit. More advanced and settled cultivators, too, often claim the sole right to gather a natural harvest of berries or nuts from woods and wastes several miles from their villages.

It is, however, both risky and inconvenient to rely on wild trees and bushes for one's food supply; so, at a very early stage in the history of agriculture, people began to cultivate fruit-bearing trees and shrubs, raising them from seed. They did this in enclosures that they built near their homes, surrounding the plots with fences to keep out cattle, sheep, and goats that would harm the young trees. Small orchards, attached to farms or forming parts of gardens, are still found in every country; but with the growth of transport and greater knowledge of fruit-growing, large commercial orchards have been developed in countries and districts where climate and soil are particularly favourable.

Tree fruits were among man's earliest food sources. At first, tribespeople claimed rights in wild-growing groves; later, they domesticated many species in tribal orchards and plots. In the Egyptian wall painting (above) fig gatherers share their harvest with apes. Right: French apple orchard in the 15th century. The apple has been the most important fruit of the temperate zone for several thousand years.

I ay dit par des
sus ou second
livre en gene
ral plusieurs
choses des ar
bres quant je parloye de la
nature des arbres et des
plantes / et des choses com
munes apertenans a la
bouraige de chascune ma
niere de champs. Mais
a present en ce quint livre
je vueil traittier de chascun
arbre par soy. Et pour
ce que aulcunes choses sont
communes a tous arbres
et aulcunes propres je vueil
parler au premier en sermon
general du labouraige
de chascun en commun et
puis je diray du labourai
ge de chascun arbre qui est
trouve en noz contrees se
lon lordre de l. A. b. c. Affin

Nuts and fruits together can supply all the substances man needs for life and growth. They contain all the sugars, fats, proteins, mineral salts, and vitamins he requires to keep him in full health.

Nuts and tree seeds are dry, rich in fats, and protected by a tough coating; they can be stored in good condition indefinitely. But fruits, being fragile and holding both sugar and water, are difficult to store, because they are quickly attacked by the bacteria and fungi that cause decay. In the tropics, fruits of one kind or another ripen at every season of the year, but in temperate lands a fresh, ripe crop can be gathered only in the autumn. So, to supply people's needs all the year round, two great industries have developed. One is concerned with the preservation of fruit in a variety of ways—as dried fruit, jam, bottled fruit, canned fruit, fruit juice, wine, and spirits. The other provides the transportation of fresh fruit from warmer countries, or from temperate countries on the far side of the equator where a crop has just ripened.

How Fruits Help Plants

Now let us look at the other side of the picture. Why do certain trees and bushes bear soft and juicy fruits? They build up these tissues around their seeds, though it is the seed alone that can nourish their offspring, the seedling plant of the next generation. Much food that might have been used to form a larger seed, or more small seeds, goes into the soft tissue of the fruit, and this soft tissue provides food only for a man, a beast, or a bird. The biological justification for this apparent waste of the plant's food resources is that, by offering an attractive bait to animals, the fruit ensures the survival of its species. Animals take the fruit, eat the palatable soft tissues, and discard the hard seeds, which are then able to strike root. The seeds of many species of fruit can pass undamaged through the digestive systems of animals and, when they are excreted, will exploit the animals' faeces as manure.

Nuts, of course, cannot survive in this way. They are whole seeds, and any animal or bird that eats them destroys them completely. Yet nut-eating creatures spread them quite effectively by chance means. Pigs bury some chestnuts whilst rooting in the ground to secure others. Squirrels and chipmunks store nuts underground and do not always return to collect them. Birds often drop nuts while in flight and fail to find them again, or scatter and bury some while they scratch on the ground for others.

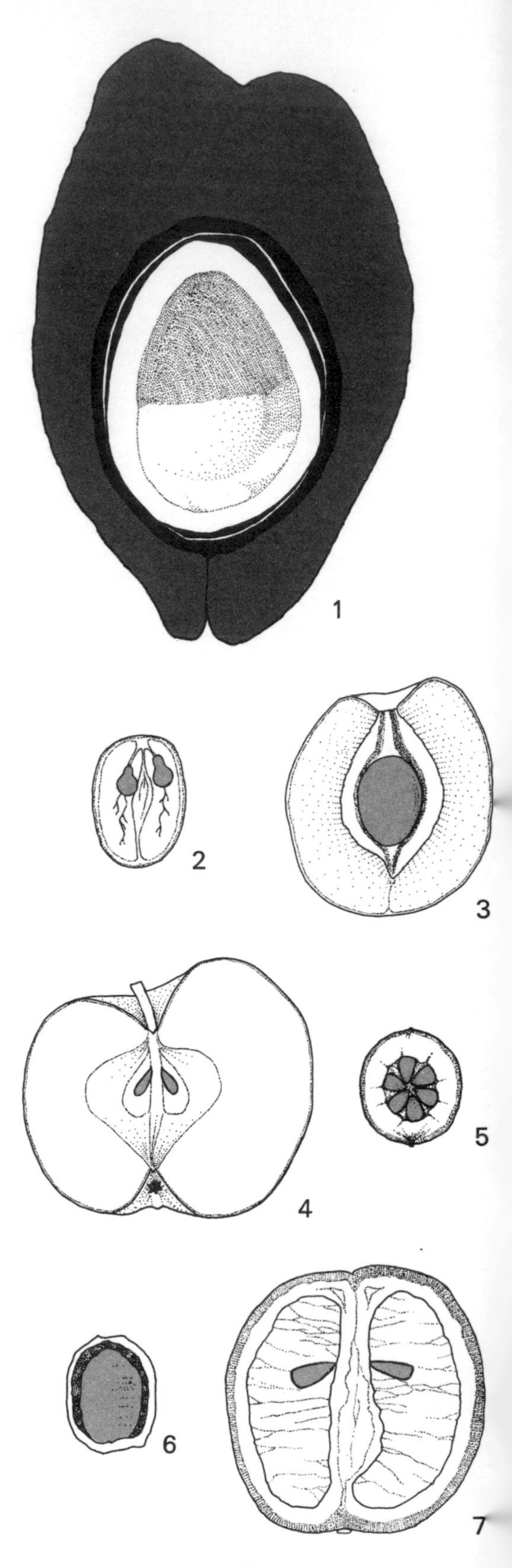

As these drawings show, man eats the seeds of nuts and the edible tissues enclosing seeds of fruits. (1) Coco-nut, (2) grape, (3) peach, (4) apple, (5) gooseberry, (6) hazel-nut, (7) orange.

Tending Fruit Trees

When man began to cultivate fruit-trees and bushes in orchards and vineyards, he had to adapt his own existence to their long life-cycle. Woody plants do not bear fruit or nuts until they have had several years in which to grow tall and branch out. Fruit-growing, then, is a mark of settled life: people plant fruit-trees or nut-trees only if they expect to stay more or less permanently in the same spot. But once established, orchards and vineyards are a valuable asset; the trees and bushes, as well as the land itself, are passed on from father to son.

Fruit-trees and bushes continue to yield their crops for many years, but the size of the crop varies greatly from one year to the next, and may even fail entirely for a season or two. During a warm, sunny summer, fruit-trees build up reserves of the sugars and other nutrients they need to form a heavy fruit crop. But none of this food goes immediately into the fruit crop: it is used to form abundant flower buds that give rise—all being well—to a heavy crop *more than one year later*. During the intervening winter, the tree's food reserves are stored in its woody stem and branches. If frost damages the young flowers, however, the heavy fruit crop may not develop at all, or it may be delayed until the following year.

Because he plans to harvest a long succession of crops, the grower finds it worth while to *graft* his trees. Grafting involves taking a shoot, called a *scion*, from a desirable tree, and causing it to grow on the main stem, or *stock*, of a less valuable tree of the same or some closely related species. It is a skilful art that has been practised from classical times onward. Grafting enables the grower to increase the numbers of a good variety very rapidly, with the certain knowledge that each new tree will possess all the desirable qualities—notably size, colour, and flavour of fruit—valued in the scion. Because scions are always taken from mature, adult wood, they bear fruit within a year or two of grafting, and they grow quickly because they have an established stock to nourish them. The stocks are raised either from seed or from suckers or root shoots of carefully chosen samples of the species.

Fruit-trees always tend to grow bulkier as well as taller, whereas vines tend to spread out farther and farther in a horizontal direction even when trained up artificial supports. Growers must prune both trees and vines every year, taking care that they do not damage the buds or side shoots that will bear the flowers for next year's crop.

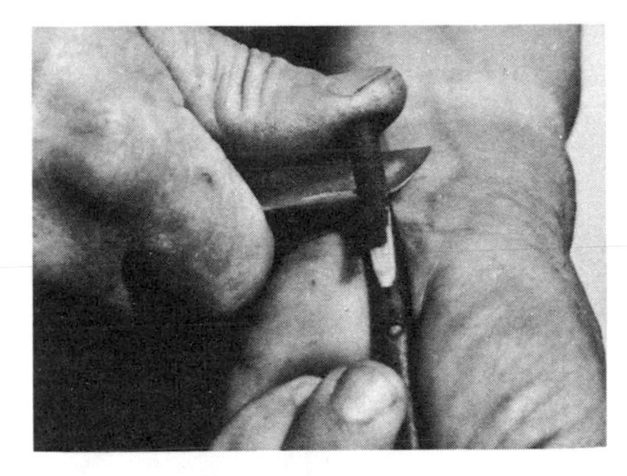

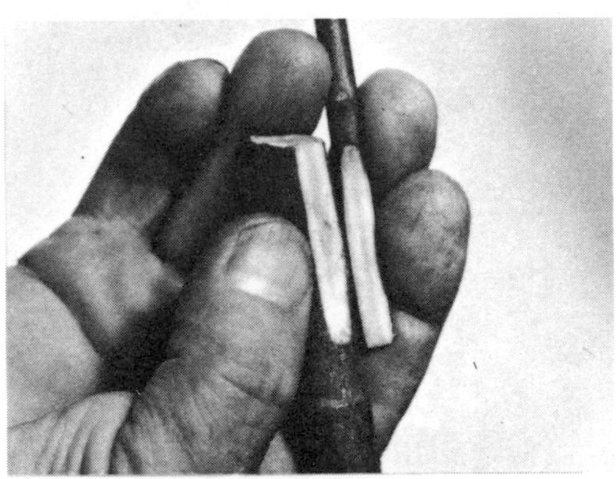

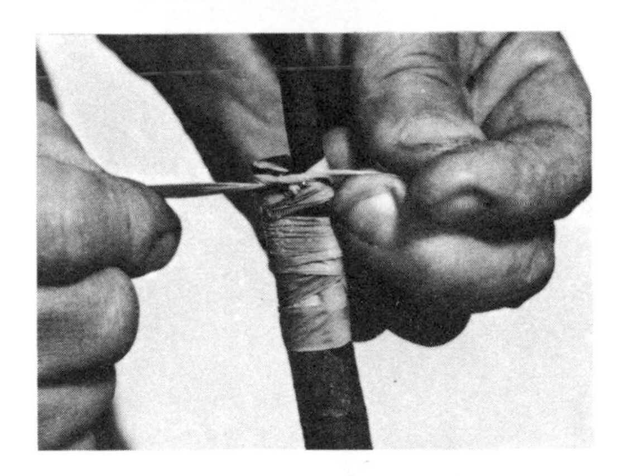

Grafting a scion from a high-yielding apple-tree onto a root-stock. The bark is stripped away and the two matching surfaces are joined, bound with raffia, and sealed with a protective wax.

The fleshy pulp of the durian, with a flavour reminiscent of almonds, is highly regarded by Malaysians. The fruit is also grown in the Americas.

Cones and pignons of the stone pine. The pignons, gathered from wild-growing trees, make a pleasant addition to the diets of Mediterranean peoples.

Wild Fruit Groves

In some countries fruits, nuts, and edible seeds are still harvested from natural groves of wild-growing trees and bushes. In Malaysia and Indonesia, for example, the durian tree (*Durio zibethinus*) is rarely cultivated, but its huge fruits can be bought, at the right time of year, in any local market. The fruit is about eight inches in diameter and looks like a large pineapple; it holds many edible seeds (each about the size of a chestnut) embedded in a tasty though unpleasant-smelling pulp. The durian ripens at the top of a tall jungle tree and falls to the ground of its own accord. Pigs, monkeys, bears, and even tigers compete with man to eat the fallen fruits; Malays (mainly Sakai tribesmen) build temporary shelters close to the trees so as to be on the spot when the great durians come hurtling down.

In Sicily and other Mediterranean lands the stone pine (*Pinus pinea*) and in Switzerland the Swiss stone pine (*P. cembra*) both develop large, nutritious seeds, or *pignons*, within their reddish-brown cones. Local people gather these cones and expose them to the heat of the sun until their scales open and the seeds fall out. In both cases, the harvest comes from stands of pines that grow wild on the mountainsides. In parts of South America the Araucanian Indians eat the seeds of another conifer, the Chile pine. This species, commonly called the monkey-puzzle tree (because its large, needle-pointed leaves supposedly deter monkeys), produces massive seeds that are collected after the seven-inch-long cones have fallen.

Another source of wild fruits that appear in abundance every summer is the low perennial shrubs of the northern moorlands. Typical of these is the blaeberry, also called bilberry or whortleberry (*Vaccinium myrtillus*), found on the uplands of most countries in northern Europe. Its juicy blue fruits are gathered by the local people, who cook them in pies; yet it is rarely cultivated.

Nuts from Northern Timber Trees

Several species of broad-leaved trees that grow wild in the woodlands, or are planted in forests for their timber, bear the large, hard, dry seeds that we call nuts. Wherever such trees grow, there is a nut crop, even if timber is the main aim of the forester. Oak-trees bear acorns, which are typical nuts, and beech-trees produce the familiar beech-nuts. Both these trees yield heavy crops only at intervals of a few years. Acorns and beech-nuts are too sour for people to eat, but many animals, especially pigs, relish them. When

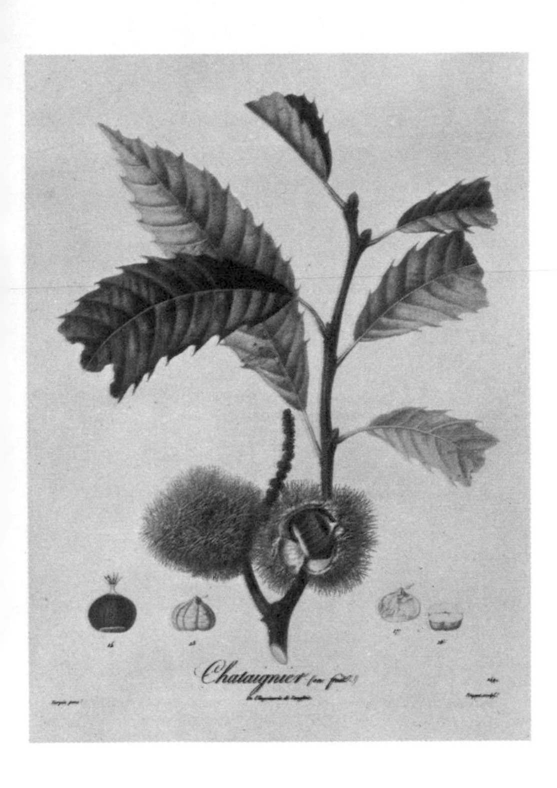

The sweet chestnut is much esteemed in Europe. Edible nuts of related species of wild-growing trees are popular in eastern Asia and in Australia.

The fruit of the walnut, like that of many other nut-bearing species, is a useful bonus to the tree's primary role as a source of high-grade timber.

William the Conqueror compiled his *Domesday Book*, a survey of all the land in England, in 1086, he judged the size of the woods by the number of swine they could support. In the New Forest, in southern England, pigs are still turned loose every autumn to feed and fatten on the acorn crop, some of it from planted trees but much from wild-growing ones.

Other nuts that are more attractive to man are sometimes picked from timber trees, and at other times are cultivated in nut orchards. The hazel-nut, also called the filbert or cob-nut, is widely grown throughout Europe. It is often found on wild bushes (genus *Corylus* of the birch family) beneath tall-growing timber, but the commercial nuts come from planted and tended orchards in which the bushes get full sunlight.

The European sweet chestnut (genus *Castanea* of the beech family) grows wild in the Mediterranean region to the south of the Alps. Its nutritious nut is tasty only after it has been boiled or roasted. The ancient Romans ground chestnuts into a flour, called *polenta*, that is still made in Italy and neighbouring lands. Polenta was a staple ration for the Roman legions, who introduced the chestnut-tree to northern Europe, including England and Wales. In the north, however, it will only occasionally ripen a good crop of fruit, though it makes a useful timber tree.

The "European" walnut (genus *Juglans*) is an example of a timber tree that has been introduced to many lands largely because it yields a pleasant-tasting and nutritious nut. It is a native of Asia Minor, but the Romans planted it in most of the countries they colonized. Later, emigrants from Europe took it to North America, Australia, and South Africa. Some of the finest walnut crops are now raised in California and other lands with a warm climate like that of its native home. Walnut orchards, formed of selected varieties grafted on to common stocks, yield a valuable yearly harvest. The nuts ripen through a soft-shelled, green stage, at which they can be gathered for pickling, to the hard-shelled, fully ripe nut. The walnut-tree is also valued as timber and is used for fine carving and turned woodware, as well as for gun-stocks.

Nuts and Fruits from Tropical Palms

The palms are the only tall trees of the great natural class of plants called the Monocotyledons, to which grasses, lilies, and orchids belong. Most palms have no branches, and consist simply of a slender, undivided stem that ends in a great tuft of leaves, together with flowers and fruit. Their stems, though woody, do not increase in thickness as do the trunks of other trees; although some become stouter at the base, they never show annual cambial growth rings. Palms bear tufts of greenish flowers, with the male and female flowers in separate groups. Their fruits take many forms, according to the species, including nutritious nuts and also soft, juicy fruits.

The coco-nut palm (*Cocos nucifera*) grew originally on the sea beaches of the Pacific Islands. It is a beautiful, stately tree, with a slender trunk that curves slightly as it grows upward for 60 feet or so, and ends in a crown of enormous, feathery leaves. The palm's growth is continued by a slender, white, pointed bud, which gives rise to fresh leaves at intervals. This bud is pleasantly flavoured and good to eat, but if it is removed the whole tree dies, and so in many countries it is protected by law as well as by custom. Groups of male flowers, and separate groups of female flowers, arise at intervals at the top of the palm. The female flowers ripen, all the year round, into huge fruits, and each tree may yield fifty nuts annually.

The coco-nut fruit, which is seldom seen whole in Europe or North America, has a brown, leathery, outer husk. Within this is a mass of tough, coarse, brown fibre called *coir*, and within this again lies the familiar woody-shelled coco-nut, which is really a single large seed. The thick, fibrous coir and shell act as a float to carry the nut unharmed across the seas, so that the tree has spread naturally from one island to another. Some growers, especially in India and Ceylon, harvest the coir to make mats, baskets, and coarse ropes and fabrics. The woody coco-nut itself holds a nutritious white flesh, rich in fats, and a white fluid called "coco-nut milk" that is sweet and refreshing to the taste.

The sap that nourishes the growing-point of the tree can be tapped without seriously checking its growth, but trees so treated rarely bear fruit. In the hot tropical climate the sap ferments within a few hours into a refreshing alcoholic drink called *toddy*, from which a strong spirit (*arrack*) is also distilled.

Most coco-nuts are grown in small orchards around the homesteads of peasant farmers living along the sea-shore or beside the great rivers of the East. The trees spring up quickly from whole fruit planted on the surface of moist ground, but they need six years of growth before they will bear fruit. Malay farmers often plant a tree when their children are born, so that it will provide them with coco-nuts from an early age—and for the rest of their lives.

Above: Jamaicans harvesting coco-nuts. As a fruit, coco-nuts are valued for their white "meat" and for their "milk," which is liquid endosperm. As a source of oil they have been harvested since very ancient times—probably at least as long as olives.

Below: Filipinos splitting coco-nuts.

Most coco-nuts are eaten locally, but coco-nut plantations are also established by commercial firms to produce oil, which is obtained from *copra*, the dried white flesh of the nuts. Coco-nut oil is used in the manufacture of margarine, soap, and confectionery. The plantations are usually sited on flat land near the sea, often on drained swamps. Thousands of coco-nut palms are spaced out in a regular pattern about thirty feet apart. As the nuts ripen they are harvested at intervals of several weeks.

The date-palm (*Phoenix dactylifera*) is cultivated as an orchard tree on the fringes of the deserts of Africa and Asia. It needs less water than most trees, but will thrive only beside rivers or in oases where some moisture is always available from underground springs or wells. It forms a stately tree with feathery, compound leaves, and bears an annual crop of fruits in huge clusters. Each date has a very hard cylindrical seed or stone at its centre, surrounded by a sweet and juicy pulp enclosed in a tough brown skin. Dates are rich in sugar and, if dried, will keep indefinitely.

Date-palms are sometimes raised from seed, but more usually from offshoots that arise at the foot of the trunk. These are carefully rooted, and then transplanted. The palms start to flower when four years old, and the Arab growers take clusters of male flowers and dust the female flowers with them, to make certain the latter are pollinated. Bunches of dates are harvested with sharp knives by men who climb the trees. Each tree may yield 100 pounds of dates a year.

The oil-palm (*Elaeis guineensis*), which grows wild in the jungles of West Africa, is a short tree that, unlike the coco-nut palm, grows markedly stouter at the base as it gets older, forming a squat, tapering trunk. At the tip of this it bears big, feathery, compound leaves and separate clusters of male and female flowers. The female flowers ripen steadily, all the year round, into bunches of fruits, each of which consists of an oily seed embedded in a thick, soft, pulpy coat. The seeds are often gathered from wild trees, and the palm is also grown in small orchards by African peasant farmers. But the most significant commercial source of the oil is the palms now cultivated in large plantations in West Africa and tropical Asia. The trees are planted about 30 feet apart and start to yield fruit three years later. Bunches of fruit are carried to the estate factory, where machines mash them up and force out the oil. About one ton of oil is obtained from an acre of palms each year. It is used in the manufacture of margarine, soap, and candles, and as a cooking fat and lubricant.

Above: cutting the leaves of the oil-palm for thatching in western Nigeria—one of the many uses for the African species in addition to its fruit oil.

Below: harvesting dates in Iraq. Each mature tree yields an annual harvest of about 100 pounds of dates, which are rich in carbohydrates and also contain fats and proteins.

In order to maintain consistently high standards of quality and yield, apple-trees are usually propagated by budding or grafting. The development of suitable root-stocks is just as important as the quality of the scions from high-grade varieties that are grafted onto them.

Left: the root-stock is allowed to grow for a year, then is cut down to soil level. New shoots grow the following spring and are encouraged to develop roots. In the summer each shoot and its root are detached from the parent stock and can then be planted out for budding.

Above, left: this root-stock has been grafted with a bud from a scion, which is inserted under the bark and secured with plastic strip. Above: in the early spring following budding, the top of the root-stock is cut off to stimulate the bud graft to develop a shoot. The shoot may grow four feet in its first year.

Left: the union of scion and root-stock is visible as a bulge near the base of the trunk. The joint should be at least six inches above the ground to protect the often susceptible scion against fungi splashed from the soil by rain.

These two apple-trees are the same age but of vastly different size—the result of bud-grafting onto different types of root-stock. Today apple-growers tend to select low-growing trees, which are easier to harvest. The root-stock also influences the age at which a tree comes into bearing and, to some extent, the size of the individual fruits.

Some apple-tree diseases are caused by viruses that can spread to all parts of the plant through the sap. This tree has rubbery-wood disease, which weakens the branches of the younger wood and makes them droop. Chemical sprays have little effect on such diseases: once infected, the tree remains so for life.

Below: this Cox's apple-tree has been grafted with a scion of the James Grieve variety, whose flowers pollinate those of the Cox's. Below, right: the variety even of established apple-trees may be changed by top-working: the main branches are cut off and two or more grafts are inserted into the ends of each stump. The tree fruits again within five years.

Hand-picking apples (upper picture) is slow and carries the risk of damaging the fruit. The harvesting machine in the lower picture is attached to the trunk of the tree and shakes off the fruit (cling peaches). The fruit lands on the elastic strips below the tree, which minimize damage on impact.

Rose-Family Fruits of Temperate Lands

In the temperate zone most of the orchard fruits are borne on trees that belong to the rose family of plants, the natural order Rosaceae. The wild-growing trees are usually too small to yield timber. The flowers are always insect pollinated—hence their showy white or pink petals that make the orchards such a magnificent sight in spring. In order to ensure that abundant insects are available to pollinate the flowers during the spring, many fruit-growers keep bees. Some varieties of fruit-trees cannot be fertilized with pollen from others of their own kind. In such cases, the grower plants a few individuals of another variety of the same species whose pollen is accepted by the rest.

Rose-family fruits are of two main kinds. One is the *stone* fruit, such as the plum, cherry, or peach, which develops a single, large, hard seed surrounded by a layer of sweet, soft pulp enclosed within a firm, tough skin. The other kind, called a *pome* or *berry*, consists of a mass of sweet flesh, holding many small seeds, within a firm outer skin. Apples and pears are typical of this group. As the fruits ripen, their colours change from green to shades of yellow, red, purple, blue, or black, according to their species and variety. At the same time their flesh loses its sour taste and changes, under the influence of the sun, to the sweet and richly flavoured ripe condition. Most of these fruits must be picked and eaten as soon as they are ripe, because they deteriorate rapidly. Some kinds of apples and pears, however, can be stored for several months under controlled conditions.

All these rose-family fruits ripen in late summer or autumn; if they are to be available all the year round in large quantities they must be preserved. Some are made into jam, some are bottled in sugar syrup, and some are canned. Some, such as the plums that reach us as prunes, are dried to lessen their water content; they hold so much natural sugar that mould fungi do not grow on them in their partially dried state. The juice of apples and pears is so rich in sugar that it can be fermented to form alcoholic drinks (cider and perry) that can be stored in casks indefinitely; apple juice is also canned or bottled in its sweet and unfermented state. Cherries and plums are used to flavour spirits, such as cherry brandy and plum gin.

Garden apples (*Malus sylvestris*) are descended from various species of wild crab-apples that flourish as small trees with pink and white blossoms and spiny twigs in the woodlands of Europe and Asia. Under cultivation three main types have been developed. *Dessert* apples have bright-coloured skins and soft

At this packing station, apples are fed by a conveyor into a machine (upper picture) that washes, dries, and polishes them. After polishing, another conveyor takes the apples through an apparatus that grades them according to size and feeds them onto the circular trays (lower picture), where they are packed.

flesh that is sweet enough to be eaten raw. *Cooking* apples, usually green or yellow, have hard flesh that becomes pleasant to eat only when they have been cooked. *Cider* apples are small, hard, and sour to the taste, but yield juices that can be fermented to make richly flavoured drinks—sweet or dry as required.

Pears (*Pyrus communis*), which are all derived from the white-flowered European wild pear, another small tree with spiny branches, are likewise divided into dessert varieties, cookers, and perry pears, while some kinds are particularly suitable for canning.

Both apples and pears are very common garden trees, but the great commercial orchards are concentrated, in each country, into warm lowlands with good soil, little summer rain, and ample sunshine. In America, for example, California is a leading state; and in England, Kent, in the warm south-east, contains many orchards. Canada, Australia, and New Zealand also grow and export great quantities of apples.

Plums (*Prunus domestica*), the leading stone fruits, are cultivated in many varieties that have been developed through the cross-breeding of several wild European and American species. The fruits vary greatly in size and colour. There are greengages and golden gages, as well as purple damsons, blue-black plums, and jet-black prunes; but the flowers of every kind are white.

The common European wild cherry (*Prunus avium*) forms a handsome white-blossomed tree in woodlands, and is one of the few fruit-trees to yield a valuable timber. Cherry-wood, cut from trees a hundred feet tall and ten feet in circumference, shows a warm, greenish-gold glow over its light-brown ground colour. Cultivated cherries, which may be black, red, or yellow, have been carefully bred from this and related wild species—notably from the sour cherry (*P. cerasus*). In orchards, all need careful pruning to stop them growing upward, like forest trees.

Peaches (*Prunus persica*) and apricots (*P. armeniaca*), both of which bear pink flowers, are native to western Asia, but are now grown throughout southern Europe and in America, South Africa, and Australia. Though difficult to keep when fresh, their fruits are easily preserved by canning. The almond (*P. amygdalus*), a closely related tree, bears no attractive pulp over the hard shell of its stone, but only a tough green skin. It is cultivated for the kernel (the almond of commerce), which lies within the tough shell. Some almonds are sweet to the taste, others bitter; all are rich in fat. Almond paste, or marzipan, prepared from them is widely used in sweetmeats and confectionery.

The orange, a native of China and the Indo-Chinese peninsula, has been domesticated for thousands of years. The blood orange, shown at left, is one of several varieties of the orange species.

Citrus Fruits in the Subtropics

In the warmer lands on either side of the equator the fruits of the genus *Citrus*—the oranges, lemons, and their allies—are the main orchard crop. They were originally native to south-eastern Asia, and have since been introduced to southern Europe, North and South Africa, Australia, and the Americas. All need abundant sunshine and are killed by even a few degrees of frost. They are evergreen, and bear shiny, leathery leaves adapted to resist loss of water during hot weather. The wild kinds, but not the cultivated ones, bear spines. Citrus fruits belong to a natural plant family, the Rutaceae, quite distinct from the rose family that gives us apples and plums. A feature common to all Rutaceae is the presence of strongly flavoured and scented aromatic oils in their stems, leaves, and fruits. These oils, which provide the orange and lemon tastes, are also used in perfumery.

Citrus flowers have five small, white, waxy petals and bundles of yellow stamens. The round fruits are at first green but ripen to shades of orange or yellow. The thick outer peel, which helps to keep the inner flesh soft and juicy, contains many oil cells; it is lined with soft white tissue, which yields *pectin*, used in jam-making.

The soft, sweet pulp is contained within about ten separate segments, most of which contain seeds or pips, though some cultivated varieties are seedless. The tough outer skin makes it possible to transport citrus fruits for long distances without loss of condition, so there is an enormous export trade in oranges and lemons from subtropical regions to northern Europe and North America. Some are preserved by canning, some are used to provide juice, and some are used to make marmalade. The peel of some varieties is candied, by boiling with sugar, to form a sweetmeat. Besides being refreshing and full of sugars, the citrus fruits are rich in vitamin C.

The many kinds of citrus fruits include sweet and sour oranges, bitter yellow lemons, fresh-flavoured green limes, and sharp-tasting citrons. All originated in eastern lands and were carried first to southern Europe by Arab and Portuguese traders, and later to the European colonies in America and Australasia. The grapefruit, however, originated in the West Indies, apparently as a hybrid.

Tropical Fruits

In every tropical land there is a great diversity of fruit-bearing trees and bushes, cultivated in gardens and small orchards, and marketed locally as they ripen,

Above: lemon-harvesting on the island of Corfu. Lemons were introduced to the Mediterranean region about A.D. 900.

Below: known to the Jewish religion as the *etrog*, the citron is one of four plant species used in the synagogue service on the Feast of the Tabernacles. The silver box for the etrog is a common subject of Jewish ritual art.

Mangoes (above) on a plantation in Brazil, where the tree was introduced by early Portuguese colonizers. Although the plant grows readily from seedlings, most commercially grown mangoes are harvested from grafted varieties.

Papaya tree (below) in Cameroun. The tree bears fruit continuously throughout its short life of three or four years.

either seasonally or throughout the year. Here we can mention only a few that are in general cultivation in the tropics, and are also exported to temperate lands. Many, such as the exquisitely flavoured mangosteen of Malaya, are too perishable to travel far.

The mango (*Mangifera indica*), which resembles a large plum, has been cultivated in India for several thousand years and is now grown throughout tropical Asia, Africa, and America. The mango-tree grows very tall and bears glossy, dark-green leaves. The juicy fruits, which fall to the ground or are gathered from the lower branches, may be eaten fresh, made into preserves, or pickled (as in mango chutney) as a flavouring for curries and other oriental dishes.

The guava-tree (*Psidium guajava*), which is native to Brazil, bears evergreen leaves and attractive white blossoms. Its fruit resembles a small apple, but has pink flesh, a gritty texture, and many small seeds. The piquant flavour makes it an attractive fruit; it is cultivated in orchards in America, India, and South Africa, and much of the crop is canned for export.

The papaya or mamao-tree (*Carica papaya*) is native to Central America and the West Indies, but was taken by Portuguese voyagers to tropical Africa and India during the 16th century. It is a small, soft-stemmed tree that shoots up rapidly from seed, but lives only a few years. It bears fan-shaped, lobed leaves, and huge, soft-fleshed, yellow fruits that can be eaten fresh or canned. The plant contains a digestive enzyme, *papain*, that is used to make meat more tender and to flavour beer.

The fig (*Ficus carica*), actually a subtropical plant, is native to western Asia, where it forms a low, bushy tree with irregularly lobed leaves. The "fruits" are really swollen stalks that hold a large number of flowers, and their resulting seeds, enclosed within them. In the wild plant, an insect must enter through a small hole at the tip of the fig to effect pollination, but cultivated figs ripen without pollination.

Bush Fruits

Pleasant, edible fruits are borne on many small bushes that grow wild as an *under-storey* in natural woodlands. Such bushes are often cultivated as orchard crops, since to give of their best they need careful tending and pruning. In Europe and North America the main bush fruits include the sharp-tasting gooseberries and red currants (both of genus *Ribes*). The black currant (*Ribes nigrum*) has a strong-flavoured fruit that is used for jams and jellies, and is a very important source of vitamin C.

Red currants (above, left), like other members of the genus *Ribes*, are native to the cooler temperate zone of Europe. The red-currant bush (above) needs to be pruned rigorously every year.
The black currant (below, left) and the gooseberry (below) are related; breeders are attempting to develop a spineless gooseberry by crossing the two species.

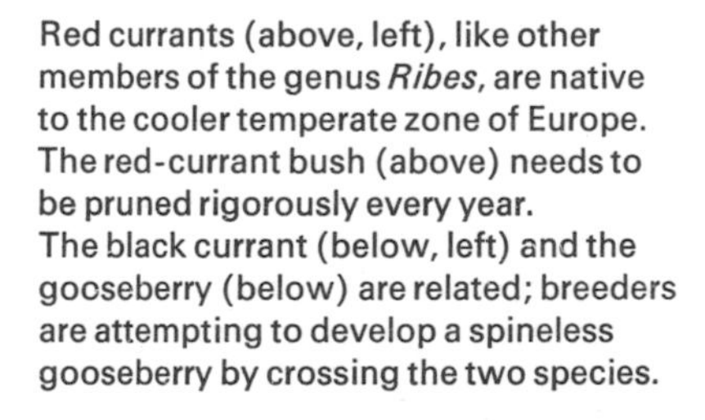

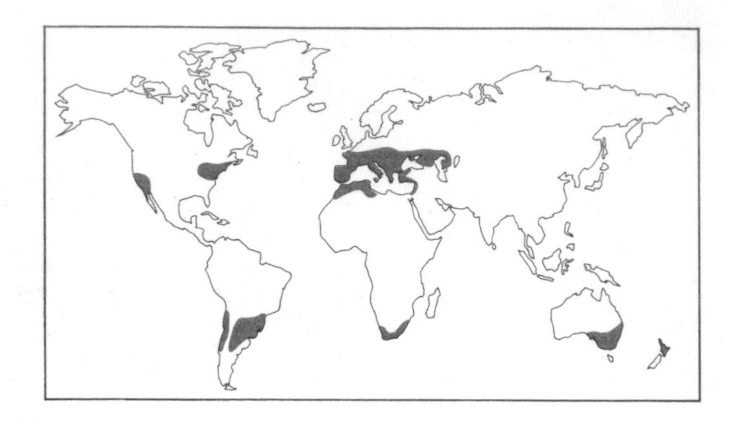

Map (left) shows the main wine-producing regions of the world.

Below: grapes of *Vitis vinifera*, the most important species in wine-making; note the bloom on the grapes' surface. Two other species that are less susceptible to diseases are grown in many vineyards in some of the hotter, damper wine-making regions of North America, Brazil, and southern Europe.

The Grape-Vine

The grape-vine is a leading source of refreshing drink and nourishment for many millions of people who live around the Mediterranean Sea and in many other lands with warm, sunny climates. It is grown in every continent, and the fruits, wines, and spirits obtained from it are eaten or drunk all over the world.

Many species of wild vines are found in Europe, Asia, and North America. They are trailing plants that live on the forest fringes, scrambling over bushes and low trees by means of tendrils. Their flowers, which are insect-pollinated, are small and attract little attention. They ripen to small, round berries, green or purple in colour (known as "white" and "black" grapes), that are borne in bunches, and each holds several small seeds. The pulp of these berries is soft, juicy, and sweet, for it holds much sugar that is readily digestible by man.

Vines need a freely draining soil. Many parts of France, Italy, and Germany that enjoy the best climate for *viticulture* are extremely hilly, and some of the finest European wines are produced from the grapes of vines that cling precariously to terraces on steeply sloping hillsides. In the wine-producing areas of the Rhine valley one can see steep hills that are dense with vines on their sunny side and almost bare on their shady side.

Vine plants are grown from cuttings or, more usually, from grafts that are first established in greenhouses, and later in nursery plots, before being transplanted. In the vineyard they are planted in rows about six feet apart and are trained to grow up trellises of poles and wires. The vine is pruned twice a year, in winter and summer. Pruning limits, rather than increases, the yield of grapes; but it has the advantage of ensuring the growth of a steady quantity of grapes of a uniform quality. A well-tended vine will

usually produce a heavy crop of grapes of medium quality in its fifth year, maintain a peak for both quantity and quality from its eighth to its fifteenth year, and continue to yield worthwhile quantities of acceptable grapes for another ten to twenty years.

All the world's great wines—and most of the lesser ones—are made from grapes of the species *Vitis vinifera*. Different wines owe their wide variations in taste, aroma, and bouquet mainly to local differences in climate, soil, and processing techniques. But the variations are also due in part to the immense number of varieties of *V. vinifera* that have been bred in different parts of the world; there are probably at least 5000 varieties (yielding black or white grapes), and most of them have been developed to thrive under specific local conditions. In France, for instance, the best clarets are made from the black grape of the Cabernet Sauvignon variety; the best burgundies from the black Pinot Noir grape; the best sauternes from the white Semillion grape; champagne from the white Pinot Chardonnay and other grapes; and cognac from the white Folle Blanche and other grapes.

The finest wines are produced in regions of the world where the combination of four factors—temperature, hours of sunlight, rainfall, and soil—provides ideal conditions for the development of the vine and the growth and ripening of the grape. The most critical of these factors is temperature. The ideal average *annual* temperature is 60°F (about 16°C) though there is a tolerance of about 8°F (4½°C) above and below this temperature. However, many regions whose annual temperature falls within this range are unsuitable for the wine grape because they are prone to frosts during the late spring or because their summers are too humid. Contrary to popular belief, very hot summers do not produce the best grapes, for although high temperatures increase the sugar content of the grape, they lower its acid content, and so deprive the wine of many of its more subtle flavours and colours. High humidity encourages mould diseases that rot the grape before it is fully ripe.

Grapes growing in the vineyard develop a delicate bloom on the surface of their skins. The bloom is somewhat sticky and it collects many millions of bacteria that are floating about in the atmosphere. Among these bacteria there are also hundreds of thousands of wild-yeast cells, including varieties that are capable of fermenting the sugars in the grapes into alcohol.

In the Northern Hemisphere, the grapes undergo their most rapid development during July and August.

Above: gathering in the grape harvest in the Charente Maritime, western France.

Below: vineyards and wine-making are central to the culture of many rural areas of France, and whole families of peasants, like this Bordelais, have worked in the vineyards for generations.

Vineyards at Kochen, near Koblenz, in the valley of the Mosel River, one of the great wine-making areas of Germany.

The amount of food manufactured by the leaves falls off markedly toward the end of August, but the grapes continue to receive water and nutrients from the roots for another three or four weeks before they are harvested. The precise moment for harvesting the grapes is critical, especially in the warmer countries where ripening is more rapid. Once the grapes have been picked they must be started on the wine-making process as quickly as possible to prevent them losing moisture. At the winery they enter a machine that crushes their skins, releasing the juice, pulp, and seeds, which are known collectively as *must*. If white wine is being made, only the juice passes to the fermenting vat. In the case of red wine, the juice, pulp, seeds, and skin (which gives the wine its colour) are all fermented together. For rosé wines the skin and seeds are removed about 24 hours after fermentation has begun.

Until the late-19th century most wine-makers relied on the wild yeasts present on the grape skins to ferment the grape sugars. Today, however, fermentation is carried out by yeasts especially grown by the wine-maker; as a result, he is better able to regulate the rate of fermentation and so can produce a better and more uniform quality of wine. Before adding these yeasts to the fermentation vat he stops the growth of the wild yeasts by treating the must with sulphur dioxide (SO_2); this chemical also raises the acidity of the wine and protects the must against attack by harmful bacteria.

The must is allowed to ferment for a period that varies from several days to several weeks, depending on the temperature due to fermentation, the sugar content of the grapes, and the type of wine that is being made. The higher the fermentation temperature, the shorter is the period spent in the vat; for red wines the temperature is usually kept below 85°F (about 29°C), for white wines, below 60°F (about 16°C). During the early stages of fermentation, the surface of the must becomes a mass of white, frothy bubbles as carbon dioxide is released. (The fizz of sparkling wines, such as champagne, derives from a second fermentation. A certain quantity of sugar and yeast is added to a dry white wine in a sealed vat, so that the carbon dioxide remains trapped.)

When fermentation has been allowed to continue to the required point, the juice in the vat is separated from the pulp and other solid matter and, in a vintage year, will eventually mature into an exceptionally good wine. The pulp is fed into a press (nowadays usually a hydraulic-ram press) that gently squeezes

Carrying the grape harvest to a winery in Portugal. Picking and transportation of the grapes are still done by hand in most European vineyards.

out the considerable quantity of juice remaining in the tissues. The wine eventually produced from this pulp juice is used mainly for blending with other, non-vintage wines.

The juice now passes through a series of settling vats, where fermentation continues at a slower rate and where various "fining" processes remove impurities—yeast cells, suspended particles of pulp, skin, and seeds, and so on—by causing them to coagulate and sink to the bottom of the vat. Here, too, begins a separate fermentation process that, in a vintage year, helps to transform a good wine into a great one. The two principal acids in the grape are tartaric acid (mainly potassium bitartrate) and malic acid. Much of the tartaric is precipitated out of the wine as cream of tartar (at one time the main commercial source of this chemical, used in baking, was the *lees*, or sediments, in the bottom of wine barrels). The strong malic acid, however, is converted into the much weaker lactic acid by *Lactobacillus* bacteria, which multiply rapidly when the yeast cells break down. Thus the wine, formerly harsh to the taste, becomes smooth and mellow. The quality of the best Bordeaux wines, in particular, depends greatly on this *malo-lactic* fermentation.

When as much solid matter as possible has been removed by fining and, later, by various filtering processes, the wine passes to an aging cask, usually made of oak. Aging, like fermentation, is an immensely complicated process involving a large number of chemical reactions. In effect, however, the wine undergoes a further reduction in acidity, develops its characteristic bouquet, and acquires its distinctive final colour. One of the most important features of the aging process is oxidation of the wine by oxygen

Below: treading the grapes to release the juices (left, from a mediaeval woodcut; right, in a modern Portuguese winery). Grape-treading has given way, in most large wineries, to mechanical crushers (see diagram overleaf).

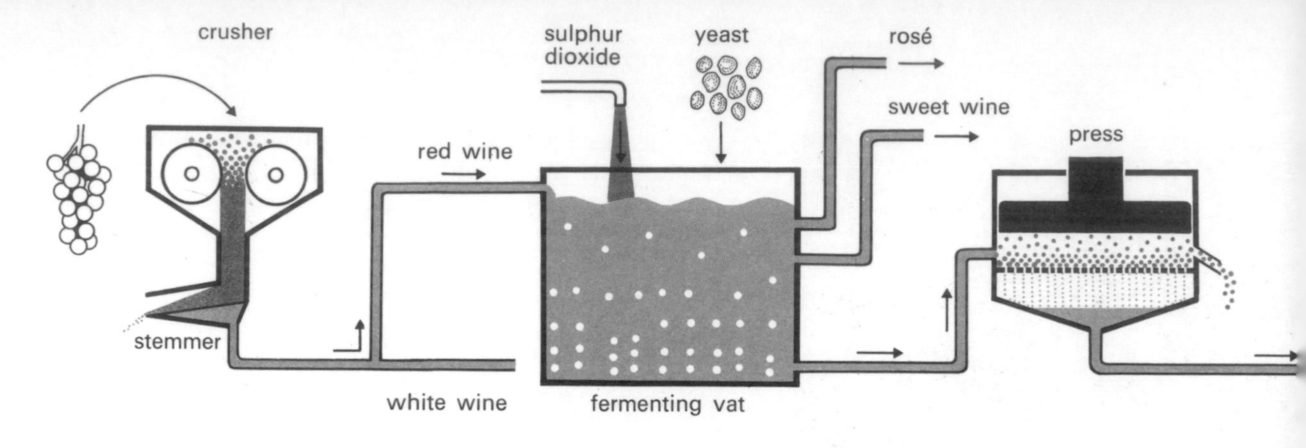

absorbed through the walls of the cask. The amount of oxidation must be carefully controlled: too little will prevent the wine from maturing adequately after it has been bottled; too much will spoil its flavour. (If exposed to air for a lengthy period, wine is attacked by the vinegar bacillus, *Mycoderma aceti*, which converts the alcohol to acetic acid.)

The term *wine* covers a wider field than is sometimes supposed. It includes both table wines (typified, for the European, by the products of Bordeaux, Burgundy, and the Rhine) and dessert wines such as muscatel, port, and sherry. While table wines never contain more than 14 per cent alcohol, dessert wines may contain 20 per cent or more, owing to the addition of brandy during fermentation.

Since brandy halts the fermentation process (owing to its high alcohol content), dessert wines contain a proportion of unconverted sugars—which explains why they are usually sweeter than most table wines. Muscatel, for instance, may contain up to 15 per cent of its original grape sugar. In the case of dry sherry, however, the brandy is added much later in the fermentation process, when the unconverted grape sugar remaining in the wine is 2 per cent or less.

Brandy itself is one of the spirits made by distilling wine; its name comes from the Dutch *brandewijn*, meaning burnt wine. The wine is heated in a closed copper still, and its alcohol (together with many pleasant-tasting substances found in it) boils off. The vapour is then condensed and stored in a closed container to prevent evaporation.

As we have mentioned, wine-making is a very ancient craft; it had flourished for thousands of years before Louis Pasteur discovered the cause of fermentation in the mid-19th century. And although the modern vintner relies as much on chemical science as on ancient custom and know-how, a great mystique still surrounds the making of wines. The fact is, an indisputably great wine is something of a freak. The vintner's art is certainly vital; but so, equally, is the

Diagram shows the principal stages in the production of wines in Europe and in California. Must from crushed grapes passes to the fermenting vat, where its sugars are converted into alcohol. Skin and seeds are removed in the press; the juice goes into settling tanks where it is purified by fining. After filtering and, in some cases, heating, it is aged in vats or casks before bottling.

Huge aging vats in the cellar of a winery producing sweet, white Tokay. The grapes grow on the slopes of the Hegyalja, a volcanic ridge near the town of Tokaj in north-eastern Hungary.

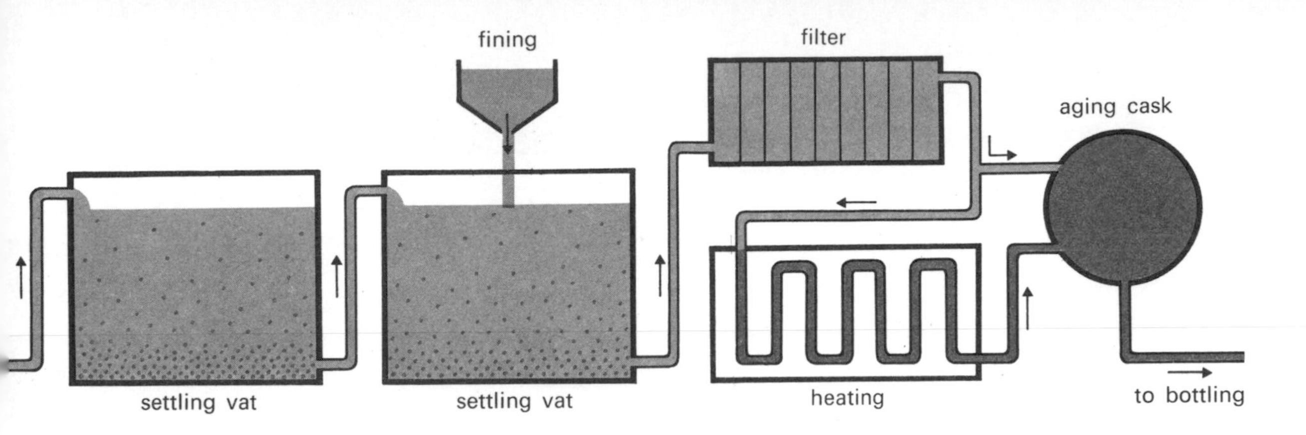

happy, unpredictable combination of ideal climatic conditions and an especially good grape crop. Much nonsense, too, is written both about the type of wine one should drink with certain types of food and about the quality of vintage wines. As to the first, snobbism has tended to conceal the obvious fact that one should be guided by one's own palate rather than someone else's. The vintage years are certainly distinguished by some excellent wines. But one should judge a wine not by the name and year on the label but by the quality of the drink itself—which, even in the best years, may vary from one bottle to the next.

The potential dangers of alcohol obscure the fact that, when taken in moderation, wine is a valuable source of energy. The amount of food we require daily in order to function efficiently is expressed in Calories. (A Calorie is a unit of heat; one Calorie—spelt with a capital C—is the amount of heat required to raise the temperature of one kilogram of water by 1°C.) Heat is a form of energy, and the caloric requirements of a man or woman vary according to the amount of energy he or she expends in a typical day. A typist, or other sedentary worker, requires about 2000 Calories; a manual worker needs about 4300. Bread yields about 67 Calories per ounce; eggs, about 46 Calories per ounce. A table wine has a value of about 20 Calories per ounce. In a warm country, such as southern France, where local wines are cheap, a manual labourer may drink at least one litre (roughly 35 oz.) of wine a day. Thus wine contributes one sixth, and possibly more, of his caloric needs. It is not only a pleasant beverage and a thirst-quencher; it is one of the most important items in his diet.

France and Italy, each producing about 2000 million gallons a year, are the major wine-making countries, and each is responsible for more than one quarter of world production. Spain and Argentina each produce more than 500 million gallons, Portugal about 400 million gallons, and the Soviet Union, United States, and Algeria about 200 million gallons.

Brandy is made by heating fine wine in a copper still (above brick furnace); the resulting vapour of alcohol and other substances is led off through the pipe and is then condensed and aged in casks.

10 Flavours, Perfumes, and Tobacco

Most of the plants we have looked at so far are necessities to the human diet; even nuts and fruit have been (and in some cases still are) vital elements in the diets of primitive peoples. But man cultivates many plants for reasons other than stark necessity. Some he values because they contain mild stimulants; some because they improve the flavour or smell of other foods; some (which we also consider in this chapter) man does not eat at all, such as perfumes and tobacco.

The stimulant properties in plants of this kind are

The spices below (see number key, right) come from different parts of plants. Cinnamon, for instance, is produced from a tree bark, vanilla from a seed pod, cloves from flower buds, turmeric and ginger from rhizomes; the others shown are seeds or fruit. Most spices must be dried and treated in other ways before they will release their essential oils.

due in most cases to the presence of *alkaloids*, nitrogenous substances that are found also in certain drugs and poisons (see next chapter). The pleasant flavours and odours derived from plants are mainly in the form of *resins* (solid or semi-solid organic substances) and *essential oils* (volatile fluids). Only rarely are the stimulants or flavours fully available in the freshly harvested plants: a coffee bean direct from the tree, for instance, has an unattractive, insipid taste. The desired properties have to be released, or strengthened, by various processes of manufacture. The processes vary greatly from one plant to another but will often include one or more of the following: (a) drying or curing to reduce water content; (b) roasting; (c) fermentation. To some extent these processes overlap. Curing, for instance, commonly entails at least some degree of fermentation; roasting, like fermentation, produces chemical changes in the tissues of the plant. Most of the products of these processes can be stored for long periods and so may be transported, without losing

1. Cinnamon
2. Vanilla
3. Cloves
4. Turmeric
5. Nutmegs
6. Black mustard
7. White mustard
8. Cumin
9. Root ginger
10. Mace
11. White pepper
12. Cayenne pepper
13. Cardamom
14. Black pepper
15. Chilies
16. Aniseed

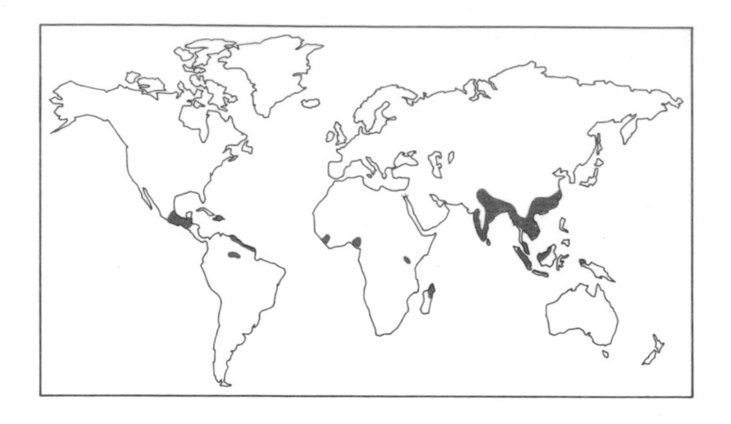

Map (left) shows the main spice-growing areas of the world. Most of them are in Asia and the export of spices to the temperate zone has played an important role in world trade since the Middle Ages.

their properties, to any part of the world. Since most of the plants involved are cultivated in the tropics and are widely used in the temperate regions, they constitute an important element in world trade. The international market for flavours has flourished for 450 years. Effective methods of keeping food, especially meat, poultry, fish, and eggs, in a fresh condition are little older than the present century. Before then food had to be preserved for the winter by drying, salting, pickling, or other methods that often destroyed much of its flavour. Moreover, even so-called fresh meat was often on the point of going bad by the time it had reached the oven. Wealthy people were quick to welcome herbs and spices, which not only imparted a pleasant flavour and fragrance but also helped to disguise the taste of food that had begun to decompose.

The spice trade between tropical Asia and Europe developed in the early Middle Ages. The spices came in Chinese junks and Arab dhows from the Far East and India to ports on the Persian Gulf and Red Sea, and thence travelled overland on camels to the Mediterranean. From here they were shipped to Venice and Marseilles and conveyed overland again into the heart of Europe. It was a long and costly journey, but even so it was immensely profitable to the comparatively few merchants at the European end who had a stranglehold on trade. The possibility of breaking the monopoly of the Venetian and Arab merchants spurred Portuguese navigators, late in the 15th century, to seek a sea route to India, to Indo-China, and even to the fabled Spice Islands (the Moluccas, west of New Guinea). Vasco da Gama rounded the Cape of Good Hope in 1497 and reached India the following year. The way was open for direct European exploitation of the spices and other riches of the East. Later, the Portuguese were to be followed by the Dutch, French, and English.

The French mediaeval manuscript (below) shows Marco Polo (1254–1324) sampling peppercorns near the Malabar Coast of India. Polo was a merchant as well as an explorer and his descriptions of Asian products, such as spices, did much to stimulate European trade with the East.

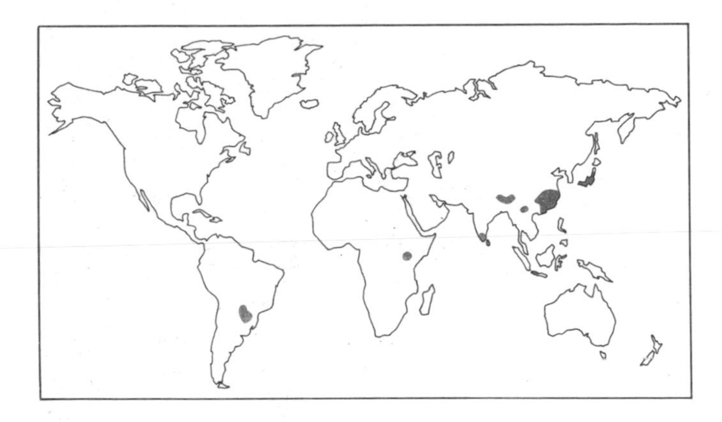

Map (left) shows the main tea-growing areas of the world. China is the largest producer, and India and Ceylon are the major exporters.

Below: the tea harvest comes mainly from the "flush"—the young growth comprising the bud and the uppermost leaves of each branch of the tea bush.

Tea

The tea plant (*Camellia sinensis*) is an evergreen bush that grows wild in the hill jungles of south-eastern Asia. It is a hardy plant and can stand frost, but it grows rapidly only in hot countries with a heavy rainfall. Under a monsoon climate it produces fresh flushes of green leaves, during each rainy spell, almost the whole year round, but in countries with a dry or cold spell its growth is more seasonal. If left unpruned it bears pretty white flowers shaped like those of the camellia (a related member of the Theaceae, or tea family). These ripen into berries containing several large seeds. Tribesmen living in southern China discovered, several thousand years ago, that if they picked tea leaves and left them to wither, the leaves developed an attractive taste and yielded a stimulating drink. We know now that the stimulant in tea is *theine*, an alkaloid similar to the *caffeine* in coffee. They began to tend the plant, and different varieties are now cultivated in many parts of China, Japan, India, Malaysia, Indonesia, and Ceylon. There are also tea plantations in East Africa, the Caucasus mountains of southern U.S.S.R., and Brazil. The English word "tea," which was originally pronounced "tay," comes from the Chinese word *te*, of the Amoy dialect. (The slang term "char" has an equally respectable derivation—*ch'a* of the Cantonese dialect.) The tea plant is quite easy to cultivate, but there are many fine details of care needed to give a high-quality crop. First the land is cleared and, in hilly country, terraced. Often trees are planted to keep the strong sun off the shade-loving tea plants. The tea plants are raised from large seeds in a nursery, and planted out in rows three or four feet apart. After three years or so the tea bushes are tall enough for plucking to commence, and from then on they will yield several harvests for many years. The harvesting of tea consists of a

Above: harvesting tea in Ceylon. During the plucking season the young leaves flush every seven or eight days.

Below: on reaching the factory the tea is withered on racks for up to 24 hours prior to rolling and fermenting. The racks, made of netting, allow the air to circulate through the leaves, so removing most of their moisture.

continuous plucking of the new shoots, so that each row of bushes resembles a low, clipped hedge about three feet high. This is usually done by hand, though clipping machines have been developed. In Assam the tea sprouts again so quickly that two Indian women must work full time to harvest a single acre of plantation. It is thus a very steady and profitable crop; productivity can be maintained for many years, provided fertilizers are occasionally applied and the bushes pruned. Labour is often recruited from a distance; the tea plantations of Ceylon, for instance, are tended mainly by Tamils from southern India.

The harvest, which consists of young, barely opened leaves and their neighbouring buds, is carried in baskets to sheds at the factory, where it is dried under cover, so that it withers. The withered leaf is then rolled, using gentle pressure, in order to release the juices within the plant cells.

The next stage in production determines whether the leaf is to be black (Indian), green (China), or *oolong* (a sort of half-way stage between Indian and China). In the case of Indian, the leaf is allowed to ferment for about 36 hours after rolling; this helps it to develop its distinctive flavour and aroma. The China leaf is steamed (but not allowed to ferment) and so retains its original green colour. The oolong leaf is

Above: an English tea-party (from an 18th-century painting). An early champion of the beverage was Dr. Samuel Johnson, who declared himself "a hardened and shameless tea-drinker."

Tastes in tea-drinking vary greatly; the use of milk, popular in England, is despised elsewhere. Below are lemon tea (left), china tea (centre), and mint tea.

allowed to ferment partially. All leaves are then dried quickly over applied heat.

The dried, prepared leaf will keep indefinitely. The tea drink is made simply by pouring boiling water on to the prepared leaf. This dissolves the theine and most of the flavouring substances. Methods of preparation vary from one country to another. In China and Japan green tea is preferred, but Russians like a strong, dark tea flavoured with lemon. In Britain, milk and sugar are usually added, and the drink thus becomes nourishing as well as refreshing. In the United States, iced tea is a popular summer drink.

When tea was grown only in China and the neighbouring lands, it was sent overland to Europe on camel trains across the Gobi Desert. Later, fast-sailing ships called "clippers" raced from the China ports around the Cape of Good Hope to bring tea to Britain and America. British settlers in Ceylon and Assam found that the tea bush grew well there, and big plantation companies were formed during the 1830s in Assam and about 50 years later in Ceylon. Today the major exporters of tea are India, which produces about 350,000 tons a year, Ceylon (220,000 tons), Japan (82,000 tons), and Indonesia (38,000 tons). World production (excluding China, the largest producer) is over a million tons a year. Britain is by far the largest importer.

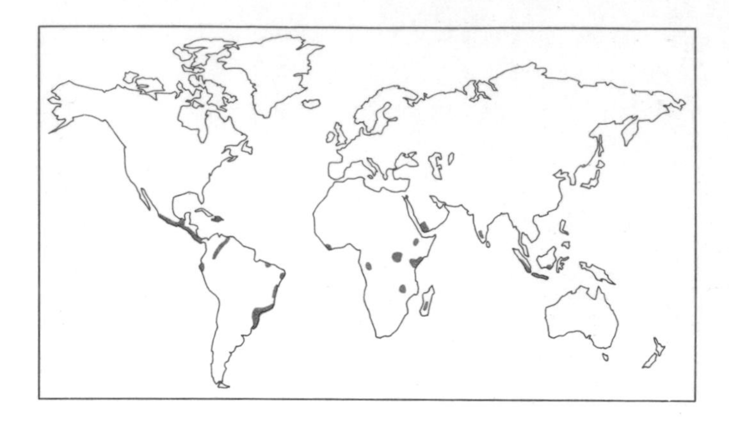

Coffee

The coffee tree (*Coffea arabica*), which may attain a height of 30 feet, grows wild on the fringes of the deserts of Arabia, Ethiopia, and Somaliland. (A somewhat less important species, *C. robusta*, is a native of the Congo region.) Its evergreen leaves have a hard, glossy surface that resists the escape of moisture during hot, dry weather. It bears, all the year round, attractive, but short-lived, white, waxy flowers that develop into reddish-brown, pulpy fruits resembling small, elongated cherries. They have a tough outer skin, a layer of soft pulp to attract birds, and two large, hard seeds enclosed within another skin. We call these seeds the coffee "beans," though they are not related to the true beans and account for only one sixth of the weight of the whole coffee fruit.

Under cultivation, coffee trees are raised from seed. The seedlings are planted out, about 10 feet apart, on fertile, carefully weeded land. They grow rapidly and start to bear fruit only three years after planting. They must be pruned regularly to keep them low and well-shaped, and to encourage fruit-bearing side branches. In some countries the crop is seasonal, in others it is gathered at intervals throughout the year.

There are two methods of extracting the berries from the fruits. In the simplest and oldest one, the fruits are left to wither in the sun, and then the pulp and the two skins are rubbed off by hand. Nowadays most of the crop is prepared by "pulping," which involves breaking up the fruits with simple machinery under a constant stream of water. The cleaned beans are next allowed to ferment slightly for about 24 hours. Then the last traces of pulp and skins are washed away and the beans are dried in the sun, cleaned, and packed for storage or export.

To prepare coffee, the beans must first be roasted and then ground to a fine powder. Much of the subtle

Map (above, left) shows the main coffee-growing areas of the world. Coffee pickers (above) in the Nilgiri Hills near Calicut, southern India. Each small, shrub-like tree bears from eight ounces to about four pounds of coffee berries a year.

Below: after picking, the coffee berry is allowed to soak and ferment; then the skin and pulp are removed, leaving two coffee beans, whose hard cases (endocarp) are dried in the sun and machine-stripped.

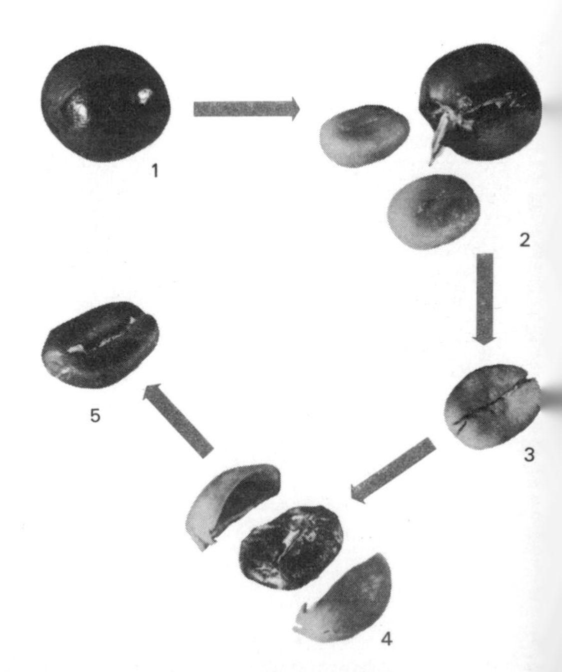

flavour of good coffee arises from volatile substances formed during the roasting, a process that must be carefully controlled if these substances are to be at their best. Ground coffee deteriorates quickly, owing to oxidation of its flavour- and aroma-producing ingredients. For this reason grinding is best done just before the coffee is needed for use. The drink is prepared by pouring very hot water on to the finely ground beans, or else by forcing steam through them, using a percolator or similar device. Only about a quarter of the ground bean's mass is dissolved in the liquid, but this fraction includes some nourishing carbohydrates as well as the caffeine and the attractive flavours.

Coffee was first used as a drink in Arabia and neighbouring lands. Arab traders carried it to France, and from the 17th century onward its popularity increased in all European countries. The coffee tree was introduced to Indonesia in the late 17th century and to Central and South America and the West Indies in the 18th century. The coffee industry of tropical Africa (based mainly on *C. robusta*) has been of world importance only since World War II. Of the present world production of some 4 million tons a year, Brazil produces 1,600,000 tons (about 40 per cent); the other major producer in the Western Hemisphere is Colombia with 470,000 tons, while Mexico, El Salvador, and Guatemala each produce more than 100,000 tons. In Africa, the largest producers are the Ivory Coast (260,000 tons) and Angola (170,000 tons), while Ethiopia and the Congolese Republic are also important exporters. Indonesia is the main Asian exporter. The United States consumes roughly half the world crop—about one third of its share in the form of "instant" coffee. The largest European importers are France and Belgium.

Above: the picked berries are placed in these tanks and soaked in water for about 24 hours; the water is then drawn off and the berries allowed to ferment for a similar period before the skin is removed.

Below: in this coffee factory, newly roasted and ground coffee is fed into cans (left). The air in the cans is removed and replaced (right) with carbon dioxide, which prevents loss of flavour.

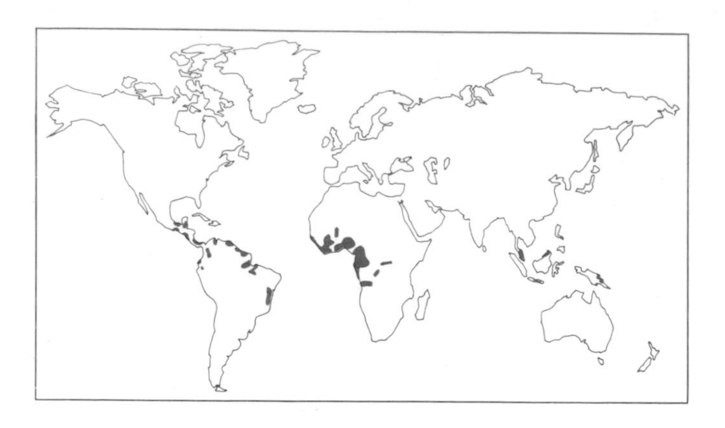

Cacao

Cacao is the collective name of several trees (mainly the species *Theobroma cacao*) that give us both cocoa for drinking and chocolate for eating. *T. cacao* was first used by the Indians of Central and South America, who found it growing wild in their tropical rain-forests. It springs up so readily from fallen seeds that nobody can now say where it originated. After his fourth voyage to America in 1502, Columbus shipped home cacao beans to Queen Isabella of Spain, and by 1528 the drink had come into use among the Spanish aristocracy. Hernando Cortez, the conqueror of Mexico, was offered cocoa to drink when he was received by the Aztec emperor Montezuma, who obtained the beans as a form of tax from the peoples of the moist jungles of Chiapas, on Mexico's south-eastern border. The tree, which is easily cultivated, has now been taken to all parts of the tropics; West Africa has become the main producing region.

The cacao tree is a small one that thrives best in the shade of larger and taller trees. It cannot stand strong sun or wind, and it needs about 50 inches of rain well distributed throughout the year. Most of the crop is raised by peasant farmers, who clear the jungle, plant (or leave existing) shade trees, and set out the cacao trees about 12 feet apart. The trees are raised from seed in nurseries, and good strains are sometimes grafted on to the roots of poorer ones. They come into bearing after about three years, and from then on flower and fruit steadily—all the year round except where rainfall is seasonal—for 40 years and more, though the quality of the fruit declines after about 20 years. Although in the wild it may grow to a height of 35 feet, the cultivated cacao tree is usually pruned to 20 feet or less so that the fruit can be gathered more easily.

The pink cacao flowers are unusual in being borne

Map (above, left) shows main cocoa-growing areas of the world. Of present world production of about 900,000 tons more than half is grown in Africa; the United States and U.K. import about 40 per cent of the world crop. The cocoa pod (above), about six inches long, is split to show the pulp-covered beans.

Below: harvesting cocoa pods in Ghana. Note that the pods are borne on the main stem, not at the ends of branches like most other fruits.

in bunches directly on the stem and larger branches, not on the smaller twigs. After pollination by insects, they ripen into large, soft, pod-like fruits, about a foot long, which vary in colour from yellow through brown to purple. The fruits are gathered as they ripen and are taken to preparation sheds, where they are split lengthwise and their contents scraped out. These consist of about 40 seeds, the size and shape of almonds (and, like almonds, with hard shells enclosing nut-like kernels), embedded in a mass of white or pink pulp. At this stage the kernels are bitter-tasting.

The seeds and pulp are piled in heaps, or stacked in wooden boxes, and left to ferment for several days. During fermentation most of the pulp decays, yielding a waste liquor. At the same time the seeds undergo chemical changes that make them sweet and pleasant to the taste. They are then dried in the sun.

When the seeds reach the cocoa and chocolate factory they are roasted and then the kernels are removed from their shells by crushing. The broken kernel pieces, or *nibs*, contain at least 50 per cent fat, called "cocoa butter"; they are also rich in carbohydrates and proteins, and contain *theobromine*, a stimulating alkaloid, as well as a little caffeine. Cocoa powder (drinking chocolate) is prepared by removing a little over half the fat content in filter presses. The hard cake that emerges from the presses is then pulverized and sifted to form the familiar powder. The cocoa butter, which is solid at room temperature but melts at body temperature, is widely used in the cosmetic and pharmaceutical industries. The butter is also added to the nibs in the production of chocolate confectionery. This addition of fat (usually about 16 per cent) helps to make the mixture malleable so that it may be more easily cast into bars and other shapes. Milk chocolate consists of about 42 per cent sugar and up to 22 per cent milk solids.

After the cocoa beans have been removed from the pods (above, left) they are piled into mounds, covered with plantain leaves (above), and allowed to ferment for five or six days. The fermentation improves both the flavour and keeping properties of the beans.

Below: in milk-chocolate manufacture, cocoa nibs, sugar, cocoa butter, and fresh milk are placed in a *mélangeur,* a revolving pan in which rotating rollers that both grind and stir reduce the mixture to a soft paste.

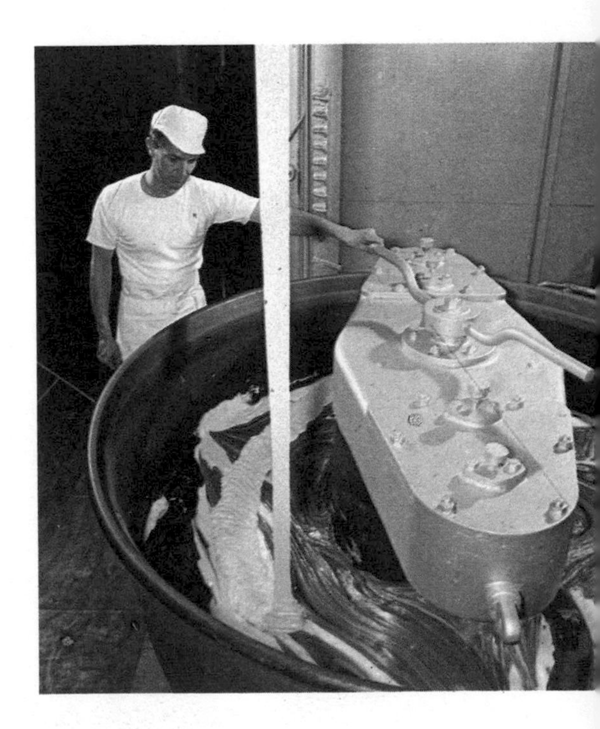

Hops

The perennial hop plant (*Humulus lupulus*) is remarkable for being grown only as a flavouring for another drink. An extract of hop fruits is used to give a bitter flavour to some, though by no means all, kinds of beer. It also improves the beer's keeping qualities.

The wild hop plant is a climber that is native to southern Europe. In cultivation it is grown as a perennial, and increased by root cuttings. Each plant is either male or female. Only the female plants yield useful hops, but a few male plants have to be included in each plantation, or "hop garden," so that their pollen will fertilize the female flowers. Hops need rich soil, liberal manuring, and a sunny climate. Their climbing stems are trained up strings suspended from wires, strung between stout poles about 15 feet high. The hop plants continue to fruit for 15 years or so, and only the strings need annual renewal.

In late spring the hop stems shoot rapidly up the strings, twining around them as they grow. They bear side branches and bunches of greenish-yellow flowers, which are pollinated by the wind. The female flowers ripen to clusters of curious open fruits. In September, the strings are cut down and the whole stem falls to the ground. The hop fruits are then picked off, nowadays usually by machine. The fruits, which look like clusters of open green leaves, are taken to a drying shed or "oast-house" and withered rapidly by hot air. The dried hops are stored until the beer brewer needs them. All he gets from them is a very small amount of flavouring substance, obtained by washing out with hot water. The "spent" hops, which hold much of the nutrients fed to the crop, are used as manure.

Irrigation pipes (above) water the young plants in this hop garden in Kent. The photograph was taken in June; within three months the plants will have more than doubled in height.

In early autumn the hop plants are cut at ground level and taken to the picking shed (left), where the female fruits are removed by machines.

The female hop fruits (below), which are like open clusters of leaves, contain a complex, bitter essential oil.

Eastern Spices

Spices are plant products that add flavour and zest to meals that could otherwise prove monotonous. They are used in many ways—as sauces, pickles, and in stews, curries, and other forms of cookery—but they cannot be eaten alone because they are too strongly flavoured. Most contain volatile oils that are released when the plants are broken up and cooked. The amount of spice used in food is partly a matter of local or national custom, and partly a reflection of the character of the main diet. For example, in those parts of India where rice is the main food, it is nearly always eaten with a curry comprising hot-tasting spices, as well as small amounts of meat, fish, or eggs.

Turmeric (*Curcuma longa*) is probably used in greater quantities than any other spice, for it is the main ingredient of the curry powders used in India and other eastern lands. It grows wild there, as a perennial, but is usually cultivated as an annual crop, increased by offshoots. It grows from rhizomes, which send up tufts of straight leaves, rather like those of an iris. The rhizomes are harvested and dried, and then ground up to give a yellow powder with a musky flavour and a strong taste; the powder is also widely used as a dye.

Ginger (*Zingiber officinale*), which grows wild in Malaysia and is cultivated in southern China, has a similar appearance and habit of growth. Its rhizomes are sometimes eaten whole, after pickling in sugar syrup, and sometimes dried for subsequent grinding to provide a sharp flavour.

Cinnamon is the dried inner bark of a small tree (*Cinnamomum zeylanicum*) grown in Ceylon and southern India. Its young shoots are cut back about twice a year, but it readily sprouts again. The bark is first scraped from the twigs by hand, and then piled in heaps to ferment. Then the outer bark is scraped away, again by hand. The remaining inner bark rolls itself up into "quills," or cylinders, that are reddish-brown in colour and aromatic. Cinnamon was one of the first spices to be brought to Europe from the East.

Cloves are the sun-dried flower buds of a small evergreen tree (*Eugenia caryophyllata*) that is cultivated on the East African islands of Pemba and Zanzibar. It is raised from seed and starts to flower five years after planting. Once a year it puts forth numerous small, green buds that ripen through pink to dark red. They are gathered just before they are due to open, and sun-dried on mats in the hot tropical sun, which makes them turn purplish black. They hold an aromatic oil, *eugenol*, that is sometimes used by dentists to deaden the pain of toothache.

Cloves being picked (above) and sun-dried (below) in Zanzibar. The clove tree grows to a height of about 50 feet and it takes skilful climbers to harvest the full crop from each tree, which may amount to 80 pounds of dry cloves. A 17th-century monopoly of the Dutch on the Moluccas Islands, cloves now come mainly from Zanzibar and Madagascar.

The black-pepper plant (above) is a perennial climbing shrub. Its small, black seeds are among the most important spices imported into the temperate zone. (From a French botanical work of 1587.)

The seed pods of cayenne pepper (below) are used to flavour curries and stews.

Black pepper (*Piper nigrum*) is grown as a vine (or climbing plant), mainly in Indonesia but also in India and Thailand. It is raised from seed and trained up living trees specially planted for the purpose, or up poles of durable wood. Each vine lives for many years and bears an annual crop of very numerous small berries, in strings like necklaces. The berry is dark red in colour, but turns black when dried. It has a soft pulp that encloses a hard, white seed, or peppercorn. Black pepper is obtained by drying and grinding the whole fruit. White pepper is made by washing away the skin and pulp just after picking, and grinding the white peppercorn, which contains a pungent resin and a number of essential oils.

Cayenne pepper consists of the ground-up seed pods and seeds of the capsicum plant, or chili (*Capsicum frutescens*); it is also called red pepper, paprika, and tabasco. The capsicum is native to tropical America, but is now grown throughout the tropics. It is an annual with a small, white flower, and bears bright-red seed pods, very hot to the taste.

Mustard, another hot-flavoured spice, is obtained by grinding up the seeds of mustard plants, several species of which are common weeds in the cornfields of Europe. Both white mustard (*Brassica alba*, the preferred species in England) and black mustard (*Brassica nigra*, which is more popular on the continent) are cultivated as annual crops. After reaping and drying in stooks, they are threshed like grain and the seeds crushed to provide mustard powder. The hot flavour does not develop until water is added to the powder.

The vanilla plant (*Vanilla planifolia*) is a climbing orchid that grows up the stems of jungle trees in tropical America, notably south-eastern Mexico. In cultivation it is increased by cuttings. Because its pretty flowers do not always set fruit, it is usually pollinated by hand. Vanilla spice is obtained from its fruit or seed pod. The fruit has no flavour until it is cured by fermenting it in a barrel for a day, then drying it for a week in the sun, followed by several weeks in the shade. Natural vanilla is expensive, and nowadays it is rarely grown because chemists can synthesize its essential flavour, *vanillin*, by the chemical treatment of wood-pulp waste.

Nutmegs are the kernels of the fruit of a small tree (*Myristica fragrans*) that is native to the Moluccas in Indonesia, and is also cultivated in the West Indies. The kernels are ground to a fine powder just before use. Mace, another spice, consists of the dried pulp, or *aril*, that surrounds the nutmeg kernel.

COMMON HERBS

Angelica

Caraway

Coriander

Marjoram

Parsley

Peppermint

Rosemary

Red Sage

Thyme

Previous page: nine commonly used herbs grown in the temperate zone.

Below: a 15th-century herb garden. At this time, monks in many parts of Europe cultivated herbs both for flavouring food and for their sometimes illusory medicinal properties.

Herbs from Northern Gardens

Before tropical spices become available at prices most people could afford, Europeans and North Americans used a variety of herbs to flavour an otherwise dull diet. With most herbs, it is the leaves and young stems that provide the attraction. These are sometimes used green, and sometimes dried for storage. Like spices, they contain resins and essential oils that modify the taste and aroma of foods.

In the Middle Ages herb gardens were tended by monks, who carried fragrant plants from one country to another. Noblemen grew herbs near their castles, and even peasants had their small plots. Today, those few herbs that survive in common use are grown mainly on herb farms or by market gardeners for local sale. They are mostly perennial plants that live for many years and send up fresh shoots each summer; they are increased by dividing their root-stocks.

Only a few natural orders of plants hold strongly-scented oils. One of these is the parsnip family, or Umbelliferae, which can be recognized by its big heads of very numerous small, white flowers, borne on radiating stalks, like the ribs of an umbrella. Parsley (*Petroselinum sativum*), which is chopped up fine and mixed with milk, flour, and butter to make a sauce for fish, is one of these herbs. Another is angelica (*Angelica archangelica*), a fragrant plant whose stems are crystallized with sugar to form a sweetmeat. Yet another is caraway (*Carum carvi*), the seeds of which give a piquant flavour to cakes.

Another natural order that includes many herbs is the mint family, or Labiatae. Most plants in this group have square stems, not round ones. Their flowers, which are often blue, show a marked division into upper and lower "lips," like the lips of the mouth. Sage (*Salvia officinalis*), thyme (*Thymus vulgaris*), and mint (genus *Mentha*) all belong to this group.

Peppermint (*Mentha piperita*), which is a good example of a herb still used commercially, is a low-growing plant, somewhat resembling the nettle, with greyish-green leaves; it grows wild in moist places in Britain, and is cultivated on thin chalk soils on the downs of south-eastern England, where it is found to yield most oil. It is increased by runners—long shoots that take root. Just before the plants are due to flower in August, they are cut with a sickle and taken to the peppermint-oil distillery. Here they are heated over water, in a copper still, and their active principle, *menthol*, evaporates and is condensed and purified. It is used to flavour mint sweetmeats, in perfumery, and in the cosmetic industry.

Perfumes

From the earliest times men and women have found pleasure in the scent of flowers. Fragrance is one of several ways in which plants attract the insects that carry pollen from one blossom to another. It comes from small quantities of volatile oils that escape into the air and can be detected by our keen sense of smell. It vanishes as soon as the flowers fade, but people like to recall it at other times and places by preserving the essential oils that produce it. The making of perfumes is a luxury trade, which satisfies a desire rather than a real need. But perfumes are needed in only small amounts, and they fetch high prices, so it can prove profitable to grow plants for this purpose.

Most perfumes marketed commercially consist of very small amounts of plant essences dissolved in a solvent, which is often an alcohol but sometimes a fat. Sometimes the flowers, or flower petals, are simply dried and preserved in closed jars. This is effective for a time with rose petals and lavender flowers, but the scent slowly loses its strength and quality.

Three main methods are used for extracting the delicate, fragrant oils from the flowers that secrete them: distillation, solvent extraction, and cold-fat extraction.

Lavender provides a good example of the distillation method. The lavender bush (*Lavandula officinalis*) grows wild in central Europe and is cultivated as a crop in England and France. It has upright stems that bear greyish-green leaves and spikes of smoke-blue

In a perfume factory at Grasse, France (above), several tons of orange blossom await processing. The yield of ***absolute,*** or concentrated flower oil, from all these blossoms will be just enough to fill the small bottle held by the foreman.

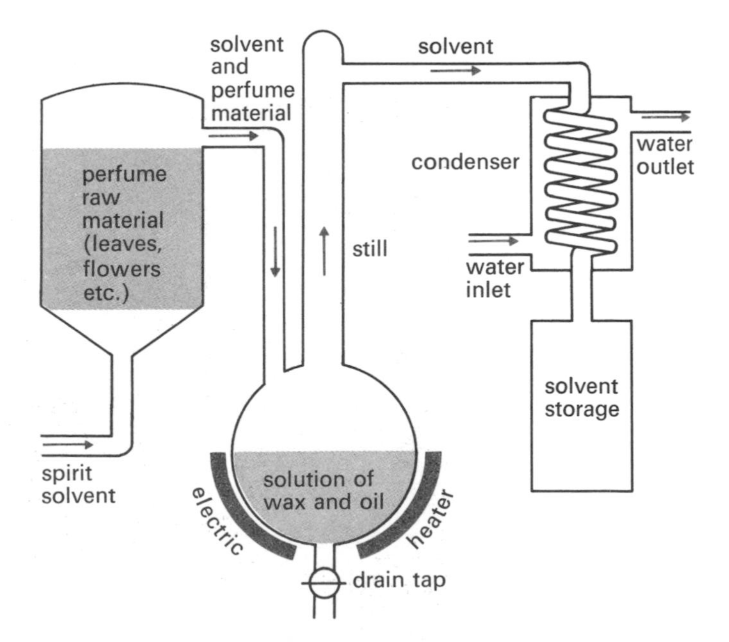

Left: solvent extraction of perfume essence. The raw materials are placed in tank at left, which is then flooded with a solvent, such as petroleum ether, that penetrates the plant tissues and dissolves their oils. The liquid then flows into the still, where the solvent is evaporated either by the application of very gentle heat, as here, or by reducing the air pressure, and passes into the storage vessel at right. What remains in the still is a waxy substance called *concrete*, from which the absolute is extracted by further processing.

A field of lavender (above, left) at Grasse; its oil is used primarily for blending with other oils. Roses (above) are in cultivation near Sofia, Bulgaria, where they are grown exclusively for their essential oils.

flowers, and the whole plant is fragrant. It belongs, like many other herbs and sweet-scented plants, to the mint family. Freshly picked leaves, young stems, and flowers are distilled over hot water, and the steam causes the volatile oil to leave the plant tissues. It is then condensed and purified for use in the cosmetic and soap industries. The process resembles that used to make some flavourings, such as peppermint oil.

Distillation is also used to prepare the scent called rose oil or attar of roses. In the Valley of Roses near Sofia, Bulgaria, millions of damask-rose bushes (*Rosa damascena*) are grown for this purpose alone. Their flowers are picked, just as they open, by women and children, from April to July. The bushes are productive for 20 years or more.

Solvent extraction is used for very delicate perfumes, such as those of the jonquil narcissus (*Narcissus jonquilla*) or the mignonette blossom (genus *Reseda*). The flowers are soaked in a chemical solvent, usually a highly volatile petroleum spirit, that dissolves their fragrant oil. The solvent is later evaporated from the mixture, and leaves behind it a tiny quantity of the heavier flower essence.

Cold-fat extraction uses the principle that fats will absorb any volatile oils with which they come into contact. For example, the flavour of butter can be spoilt by exposing it to the smelly volatile oils of fresh paint. If fragrant, fresh flowers are pressed against a fat such as lard, it will absorb the oils containing their scent, and later on it can be separated out by using

Below: a workman gently turns the rose petals in a drying shed at a Grasse factory. A ton of these petals normally yields about a pound of attar of roses.

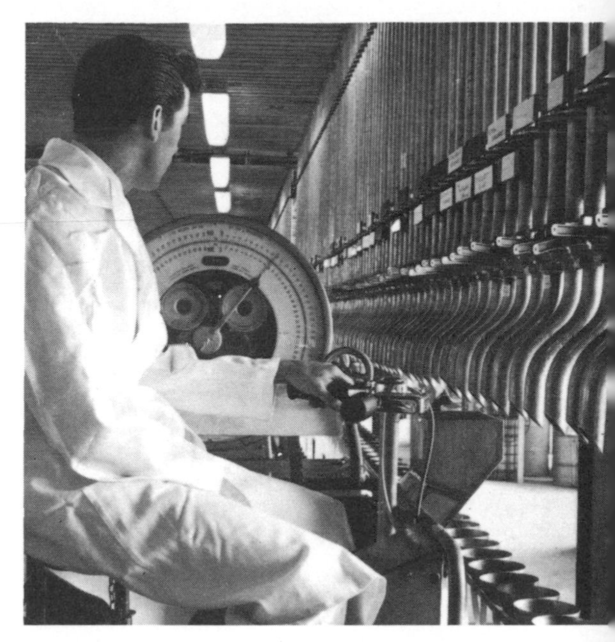

A chemist (above) develops a new perfume at an "organ," a cabinet lined with shelves containing hundreds of bottles of different essential oils. When the exact proportion of each oil has been determined, the new perfume can be made in bulk. The compounder (above, right) uses a motorized weighing machine to collect from storage tanks the essential oils needed for quantity production.

alcohol as a solvent. This method is used for the delicate scent of jasmine blossom (*Jasminum officinale*), which is grown for this purpose around Grasse, the centre of the perfume industry in southern France.

Perfumes are also obtained from other parts of plants besides the flowers. Important sources are lemon-grass leaves, geranium leaves, orange peel and lemon peel, the wood of the sandalwood tree, and certain mosses and lichens. Chemists have discovered ways to make certain scents artificially, but they can seldom reproduce exactly the natural odours of plants. Many perfumes sold commercially include artificial ingredients, with small quantities of natural scents.

Nectar

Those plants that rely on insects to carry the pollen from one flower to another reward the insects with nectar, a sweet, viscous liquid secreted by glands called *nectaries* in the heart of the flower. If one sucks the base of a honeysuckle flower, or a clover blossom, one can taste the nectar; with practice, it is possible to distinguish between the nectars of different species. But in general all the nectar we eat is harvested for us by bees, who preserve it in the form of honey. Beekeepers well know that they must place their hives near plants that flower abundantly and yield much nectar. In Europe and North America these are usually clovers, orchard fruits, or the wild heathers that yield honey of distinctive fragrance. In Australia wild eucalyptus trees are a major source.

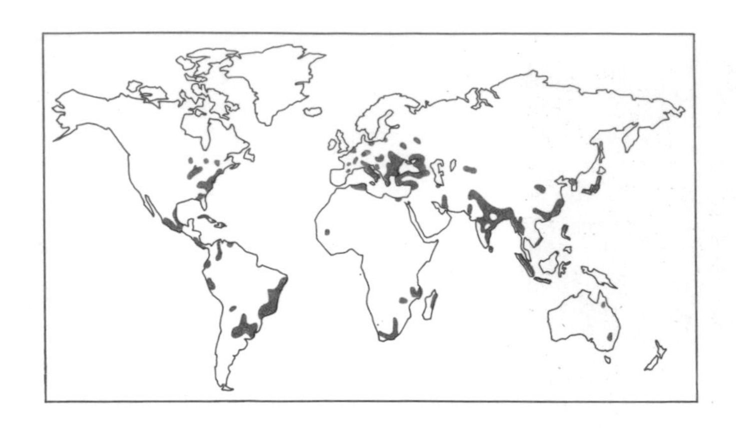

Above, left: map shows main tobacco-growing areas of the world. Above: the pipe-smoking figure (left) is from the picture-writings of the Aztecs. The English pipe-smoker (right) is from the title-page of a play by Middleton and Dekker published in 1611.

Tobacco

Whereas the caffeine in coffee and tea acts as a mild stimulant, the alkaloid in tobacco, *nicotine*, is a narcotic: in common parlance, it "calms the nerves." In large doses nicotine is poisonous (it is the basis of certain insecticides); in the amounts available in tobacco, however, it is simply habit-forming.

The wild tobacco plant (mainly the species *Nicotiana tabacum*) is a native of tropical America and is a perennial; in cultivation, however, it is grown as an annual. In the wild, it reaches a maximum height of about six feet and bears self-fertile, tube-shaped, white, pink, or red flowers. Each plant will, if allowed, develop about 200,000 tiny seeds.

Tobacco has been cultivated by the Indians of America for at least a thousand years, and probably for much longer. For many tribes tobacco had deep religious and ceremonial significance; some believed it to have medicinal properties as well. Most Indians at the time of Columbus's arrival smoked tobacco in pipes made of reeds or tubes of cane. Spanish explorers, however, introduced it into Europe, early in the 16th century, in the form of crude cigars. European settlers began to cultivate tobacco in Cuba and Brazil during the 16th century, and within a hundred years it was also being grown in what are now Virginia and Maryland. Since then its cultivation has spread to many parts of South America, Europe, central and southern Africa, western Asia, India, China, Japan, the Indo-Chinese peninsula, and Indonesia.

As this wide distribution suggests, the cultivated plant thrives in a considerable range of climates. One reason for this is the very large number of varieties developed to suit particular local conditions. In addition, a closely related species, *N. rustica*, is grown in the Soviet Union, Turkey, and parts of Europe and India. In most of the cooler countries, cultivation

Below: mature tobacco plant. In most countries the crop is harvested, not by cutting down the whole plant, but by picking off the leaves as they ripen.

begins with seedlings that have been raised in hotbeds under glass for a period of about two months. They are then planted out in thoroughly weeded, well-fertilized soil in rows about three feet apart. When the plant has reached the desired height (about three or four feet) it is *topped*: the top of the central stem, and any lateral shoots, are removed, so that growth is concentrated in the remaining leaves.

The crop is usually harvested two to four months after transplanting, and the leaves are then cured by one of several methods. Air-curing, the simplest method, is carried out in sheds that are carefully ventilated to maintain the required temperature and humidity. Another method is to cure the leaves with smoke from an open wood fire on the floor of the shed. Flue-curing involves the use of pipes around the floor of the shed. (Flue-cured leaves are what the smoker knows as Virginia tobacco, regardless of where the leaf was grown.) In all these processes, the leaves undergo very slow fermentation.

After curing, the leaves are sold to a manufacturer, who allows the tobacco to age for anything up to three years. After this it is ready to be cut, shredded, or ground, and treated in a variety of other ways to make it suitable for smoking, inhaling as snuff, or chewing. Almost all the commercial cigarette, cigar, and pipe tobaccos are blends of several varieties of leaves, each with its characteristic flavour.

Medical reports, published by scientific bodies in several countries, linking smoking (especially cigarette smoking) and lung cancer seem to have had little effect on the public demand for tobacco. One reason for this, without doubt, is the difficulty of giving up smoking once the habit has been acquired. Many people would like cigarettes to be banned. The fact is, tobacco is a double-edged social problem: unhealthy as smoking may be, tobacco is central to the export and domestic trade of many countries. In Britain, for instance, government and local authorities from time to time issue advertisements warning of the dangers of smoking; yet cigarettes provide the government with an immense revenue in the form of direct taxes (about three quarters of the retail price) that might be difficult to raise by other means.

Of an annual world tobacco production of some four and a half million tons, the United States produces about one quarter. China and India each produce almost 400,000 tons, Brazil more than 200,000, and Bulgaria, Greece, Rhodesia (all major exporters), Japan, Pakistan, the Soviet Union, and Turkey each produce more than 100,000 tons a year.

Above: hand-stripping the midribs from wrapper leaves at a Cardiff factory. The wrapper is a narrow strip of leaf wound spirally around the binder and filler leaves of a cigar.

Below: cigarette tobaccos pass from blending machines to a large silo (upper picture) where as many as 30 grades of leaf may be uniformly mixed. The machine (lower picture) manufactures about 2000 cigarettes a minute.

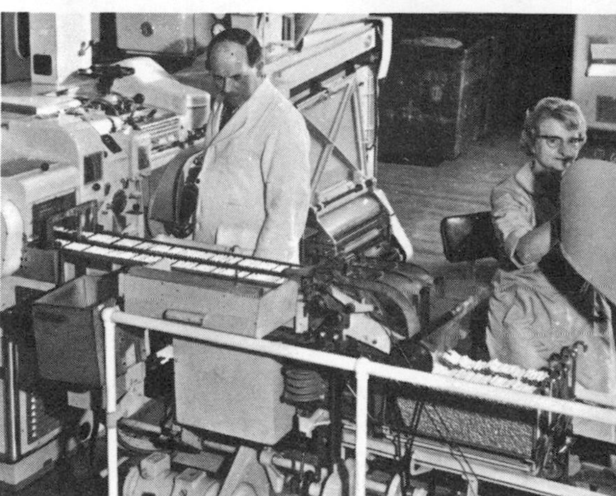

11 Drugs and Poisons

Certain plants contain organic substances that have powerful effects on the working of the body of man and other animals. Some, for instance, alter the rate at which the heart beats; others affect respiratory processes or the action of muscles; still others have the effect of deadening pain. As a consequence, these substances are extremely important in medicine, both in the treatment of disease and as aids to surgery. The interesting thing about some of them is that, although they may be life-savers in small amounts, they are deadly poisons in larger doses.

Certain of these substances, notably the pain-killers, are powerful narcotics that are capable of inducing a coma. As we saw in the previous chapter, the alkaloid nicotine in tobacco is a narcotic and is *addictive*, or habit-forming. The narcotics used in medicine are also addictive, but much more strongly so. A person addicted to morphine, for instance, suffers extreme mental and physical anguish if denied supplies of the drug. Because narcotic-containing plants are potentially very dangerous, their cultivation and use are now controlled by law in every

Below: like essential oils, narcotics and poisons are extracted from various parts of different plants. Quinine comes from cinchona bark, opium from the poppy-seed pod, belladonna from leaves and roots, strychnine from seeds, cocaine from coca leaves, pyrethrum insecticides from flower heads.

Right: production of strychnine. The seeds are first mixed with alcohol, then the liquid is evaporated in boiling pans, as here, to produce a syrupy extract. Lethal when taken in quantities greater than 50 milligrams, in smaller doses strychnine is useful in medicine.

Cinchona bark

Poppy capsules

Belladonna leaves

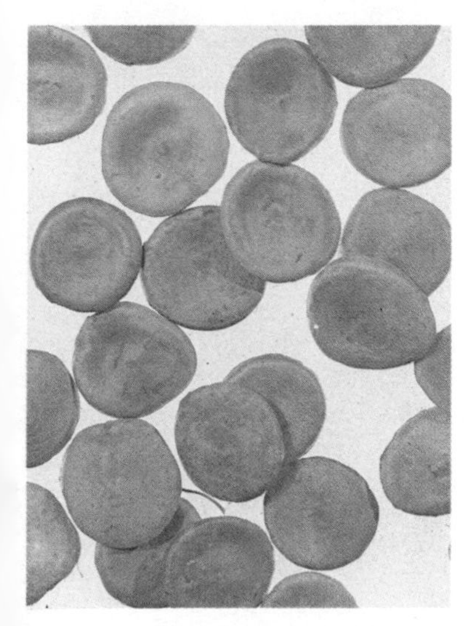

Nux-vomica seeds

Coca leaves

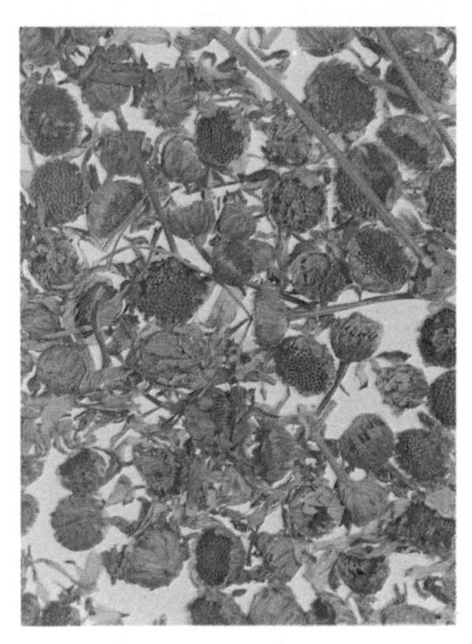
Pyrethrum flowers

country. Nonetheless, drug addiction is extremely wide-spread and is one of the most serious social problems in countries all over the world.

Knowledge of the existence of plants containing narcotics and poisons is almost as old as society itself. The lives of primitive food-gathering peoples, after all, may depend as much on this as on the ability to find nutritious ones. Certain jungle tribesmen know exactly which climbing plants will yield the poisons with which they tip their arrows—though many European experts have great difficulty in making botanical distinctions between these plants and harmless ones.

Until the modern chemical and pharmaceutical sciences identified, and revealed the exact effects of, the active principles of drug-producing plants, many strange ideas were held by herbalists and apothecaries. Some believed that if a plant resembled a part of the human anatomy it would cure disease of that part. For instance, a lichen that grows on the oak tree, and looks a little like a human lung, was once called Lungs of Oak and was often prescribed for chest ailments. Any plant with a strong, preferably unpleasant, flavour or scent was once thought to possess healing powers—if only the right disease for it could be found. We should not, however, be too sceptical about such ideas. Many credulous people have been at least partially cured by a "medicine," chemically innocuous, whose vile taste has convinced them of its efficacy.

Some of the plants in this chapter are less important than they once were because chemists have discovered ways to synthesize their active principles by other means and from other materials. Synthetic drugs are, almost invariably, more desirable than those derived from plants: they are cheaper to produce in bulk and their quality is easier to standardize. Nonetheless, as we shall see, many of the substances considered in this chapter can still be derived only from plants.

Quinine from the Cinchona Tree

Malaria has afflicted people living in humid, and especially marshy, places since prehistoric times. It is a recurrent fever that causes the sufferer to shake, renders him incapable of active work until each attack passes, and sometimes proves fatal. It is caused by a minute form of animal life called a protozoan, which can be carried from one human to another only by blood-sucking mosquitoes of the genus, *Anopheles*. Hence it can be checked to a large degree by draining the swamps in which the mosquitoes breed, or by putting oil on the surface of streams and ditches, to kill their larvae. But a number of drugs have proved

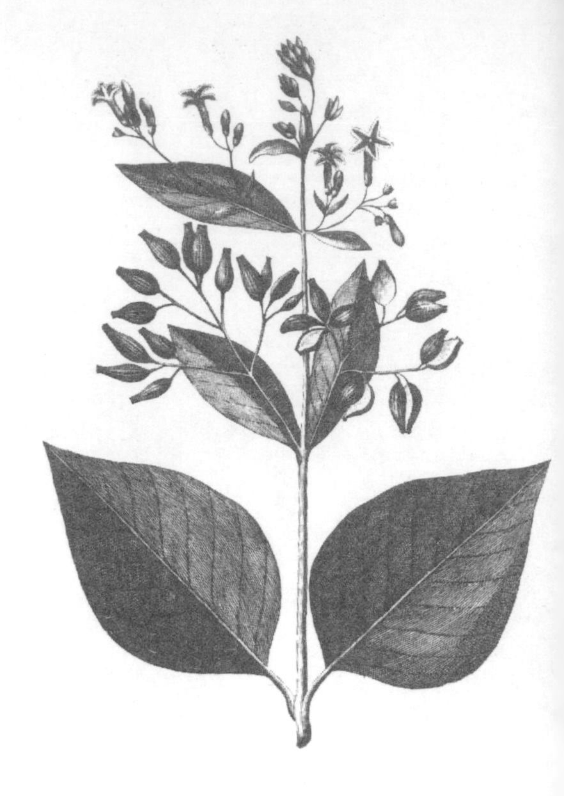

Above: leaves and flowers of the cinchona tree (from a botanical work of 1738).

Below: cinchona plantation in Guatemala; the trees are about seven years old and have been raised from root stocks to which selected buds of a high-yielding strain were grafted after six months' growth.

Cinchona bark is stripped from the trunk (above) and rapidly sun dried (below) to prevent loss of its alkaloids. In most regions harvesting involves cutting down the whole tree, for quinine can be extracted from the bark of branches and roots as well as from the trunk.

useful in curing the disease. One of these is quinine, a drug derived from the bark of various species of South American trees of the genus *Cinchona*. When this bitter drug is taken in the right amounts, it enters the bloodstream and kills the malarial organisms.

The cinchona tree, a small evergreen, grows wild in the high Andes of South America, and also in Panama. The Indians, who had used it from the earliest times, first introduced the drug to Catholic missionaries in the 17th century—hence its once-popular nickname, "Jesuit's bark." Unfortunately, the trees grew only in the remote highlands, and for long they were jealously guarded by the tribes that knew their curative property. During the 19th century, however, a British government expedition succeeded in bringing some young cinchona trees back from Ecuador. By 1860 the cinchona tree was established in India and Java, in both of which countries malaria was a serious problem.

Under cultivation the trees are first raised from seed, then grafted with a special strain that holds a high proportion of quinine in its bark. After seven years' growth they are cut down and the bark is removed and dried for shipment. The drug, which consists of four different alkaloids, is extracted from dried bark by dissolving it out with alcohol. Nowadays, synthetic drugs such as *atabrine* are also used to treat malaria, but quinine is still employed on a considerable scale. The drug is also used to prevent muscle cramp, and, in small quantities, as a flavouring for "tonic" drinks.

Foxglove—an illustration from the classic work of William Withering on the medical uses of digitalis, published in 1785.

Digitalis from the Foxglove Plant

The foxglove (*Digitalis purpurea*) is a beautiful plant that grows wild in the woods of north-western Europe. It starts life as a tiny seed that grows, in its first season, into a small tuft of leaves; the following year it develops larger leaves and sends up a tall spike bearing many thimble-shaped, purple flowers. All the material needed for drug-making could easily be gathered from the wild-growing plants, as it used to be in the past. But because chemists need a standardized product, the foxglove is now grown as a field crop in Holland. It is raised as an annual, being sown under glass, transplanted to the fields in June, and given ample fertilizer to make it form big leaves. The leaves are gathered in autumn whilst green, and dried slowly under cover in the shade. One form of the drug *digitalis* is made simply by grinding the leaves into a powder; but usually the leaves are steeped in alcohol, which dissolves out the drug to form *tincture of digitalis*. Digitalis contains several glycosides—compounds in which sugars combine with other organic chemicals. These, acting together, improve the productive work of the heart while slowing the pulse rate.

Morphine from Opium Poppies

Since the dawn of history, people have known that a powerful drug could be obtained from the seed vessels of a common poppy plant, *Papaver somniferum*. Its use was recorded by the ancient Egyptians, the Greeks, and the Romans. The opium poppy, which is probably native to Asia Minor, is found in most countries as a weed of cultivation because its tiny seed is hard to separate from that of the main-crop plants. It is often grown as an ornamental flower, and its seeds are used in confectionery and as a source of oil. Commercial crops are grown in south-eastern Europe, where cheap labour is available for the tedious work involved. The poppy, which is an annual, is raised from seed and allowed to flower. As the seed pods form, they are scratched with sharp knives, and white gum (*latex*) slowly exudes from the wounds. This is done at sunset, and the following morning the dried latex is scraped off, again by hand. One acre of opium may carry 20,000 poppy plants, yet it will yield only about 30 pounds of opium.

In medicine, opium is used mainly as a source of the drug morphine, which is extracted by complex chemical processes. Morphine is mainly used to deaden pain arising from accident or disease; it also controls coughing in diseases of the organs of breathing, and induces a feeling of euphoria, or well-being.

Opium capsule, or seed pod (above), has been scratched to make the narcotic-containing latex flow. When the latex has been collected from each capsule it is dried and kneaded into balls of crude opium, like the samples (right) seized by drug-control officers in Korea.

No satisfactory artificial substitute has yet been found for it; moreover it is the only narcotic drug that relieves pain in reasonably safe doses. But much of the opium is grown commercially for selling to drug addicts. When inhaled in smoke it causes a drowsy stupor (morphine is named after the Greek god of dreams) in which the addict first enjoys pleasant hallucinations, but later is rendered helpless for many hours. Opium-smoking is illegal in most countries, yet there are millions of people, particularly in India and China, who practise it regularly. Tincture of morphine, in which the drug is prepared as a liquid, is commonly known as *laudanum*; an artificial derivative of morphine, *heroin*, is also used medicinally and is perhaps the most dangerous of the addictive drugs.

Below: examples (left) of the opium addict's equipment—crude pipes of rubber tubing and chinaware, and cooking lamps made from tins, a salt cellar, and a tumbler. The smoker (right) places a small ball of opium into the bowl of his pipe, heats it over his cooker, and then inhales the fumes. Like other narcotics, opium must be taken in progressively larger and more dangerous quantities in order to satisfy the addict's craving.

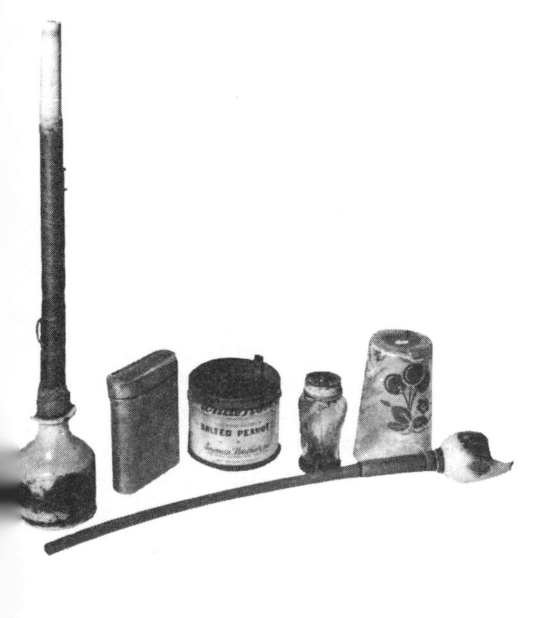

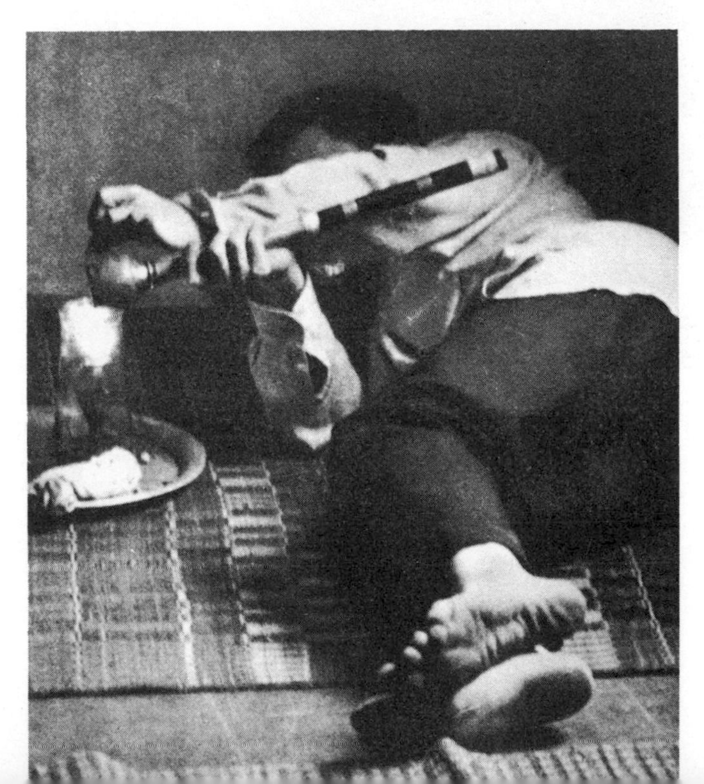

Hemp, which grows to a height of about seven feet, has separate male and female plants, hashish being derived from the leaves and flowers of the latter.

Hashish or Marijuana from Hemp

The hemp plant (*Cannabis sativa*), which has been cultivated for several thousand years as a source of fibres and of seeds rich in oil, holds a powerful narcotic resin in its compound leaves and greenish flowers. This is prepared and taken in various ways, being eaten, drunk, inhaled, or smoked; it is not used medicinally. In its strongest form, the drug is obtained from a resin that forms on the flowers of the female plant. Often the leaves are simply dried, powdered, and mixed with tobacco in cigarettes. The drug is variously known, in different parts of the world, as *hashish*, *marijuana*, *bhang*, and *ganja*. It is not an alkaloid nor, like most of the other powerful narcotic drugs, is it strongly addictive. It nevertheless poses an extremely serious social problem, for two reasons. In many parts of the world, all or most of the drugs used for illegal purposes are obtained from the same criminal sources. It is extremely common for the smoker of marijuana to move on, by choice or persuasion, to an addictive drug such as heroin and so to become "hooked" for life. Moreover, although the strains of the hemp plant that secrete the largest amounts of the narcotic resin grow mainly in India, Arabia, and North Africa, *C. sativa* is cultivated for its fibres in most countries of the northern temperate zone. Thus, potential sources of the drug are legion. The World Health Organization, however, is at present trying to develop varieties of fibre-yielding hemp that are free from the narcotic resins.

Other Leading Plant Drugs

The belladonna plant (*Atropa belladonna*), often called deadly nightshade, grows wild in woods and waste places in Europe, and is cultivated in Holland to provide the leaves and roots that yield certain potent drugs. It is a sturdy perennial that sends up a stout stem bearing oval leaves and many purple, bell-shaped flowers, followed by black berries. Chemists extract from the dried, powdered leaves, and also from the roots, two drugs called *atropine* and *hyoscine* that relax certain muscles and dry various glands; large doses, especially of atropine, cause delirium and subsequently send patients into a coma. Atropine also causes the pupil of the eye to dilate, and the plant's name—belladonna means "beautiful lady"—is due to the fact that Italian and Spanish women once used the drug to make their eyes look large and alluring.

Strychnine plants (mainly *Strychnos nux-vomica*), which yield two stimulating drugs, the bitter-tasting alkaloids *strychnine* and *brucine*, grow wild in the jungles of South America, south-east Asia, and Indonesia. They are small trees, bushes, and climbers of several related species. All bear soft fruits resembling oranges; these are gathered by the drug collectors, who break them open and extract the hard seeds for sale and export. The drugs are obtained from the seed by extraction with alcohol or acid. Strychnine, if taken in quantities greater than about 50–100 milligrams, is a deadly poison; it is used in some countries to kill harmful animals by adding it to an attractive bait. Its main use in medical practice is in awakening people out of narcotic comas; it also increases the acuteness of certain kinds of sensory perception.

Cocaine is the only widely used *local* anaesthetic of plant origin, though there are several synthetic ones. An injection of tincture of cocaine has the effect of deadening the responses of nerves in small areas of the body to pain for periods of up to one hour; it is widely used in dentistry. The drug is obtained from the coca tree (*Erythroxylon coca*), which grows wild in the Andes of South America. The tree is cultivated in South America and in Java, and the commercial drug is obtained from its dried leaves by extraction with alcohol. Indians in the high Andes chew coca leaves: in small concentrations the drug is a stimulant and improves the capacity for work. However, if used constantly in this way cocaine becomes addictive. Today cocaine has largely been replaced by synthetic compounds such as procaine, which has a similar composition and is safer.

Belladonna is derived, mainly by solvent extraction, from the leaves and roots of the deadly nightshade plant (above). The drug has been used in Europe since early mediaeval times, and until the last century was gathered from wild-growing plants. Today it is cultivated in Holland and (below) Huntingdonshire.

Arrow Poisons

Some quite primitive tribes living in the tropical jungles of South America and south-east Asia know the secrets of preparing strong plant poisons that act very quickly on human or animal prey. The poisons are used to tip their arrows and blowpipe darts. Once the poison has entered the victim's bloodstream through the wound made by the arrow, death follows within a few minutes. Several plants are known as sources of these poisons, including some species of the genera called by botanists *Strychnos*, *Derris*, and *Antiaris*. The actual method of compounding the poison is usually kept a secret, known only to the old men or wise women of the tribe.

One of the best-known of these arrow poisons is curare, made by the Indians of Guyana and Venezuela. Its most active principle, *curarine*, comes from a woody climbing plant belonging to the genus *Chondrodendron*. Sir Walter Raleigh, who sailed up the river Orinoco in 1595, was one of the first people to describe its effects: its victim, he wrote, "endureth the most insufferable torment in the world." Curare acts on the nervous system and paralyses the action of the lungs,

Above: Indian boys of the Japurá River region, Upper Amazon, poisoning the tips of blow-gun darts with curare.

Below: Indians poison fishes by beating the curare-containing juices out of a bunch of liana branches laid across a stream. The fishes are safe to eat.

so that the victim dies of suffocation. Rather surprisingly, curare also has its uses in medicine. It has been known for a century that, if given in small amounts, it can relax muscles that usually remain contracted even under anaesthetics. It is now widely used, in very small quantities, in general surgery.

Fish Poisons and Insecticides

A few plants hold substances that have the remarkable properties of stupefying fish or killing insects. One of these is the leguminous barbasco plant (*Lonchocarpus nicou*) of Peru; another is the related tuba plant of Malaysia, which is also called derris. The roots of both plants contain a toxic substance called *rotenone*. Tribesmen prepare a powder containing the poison and spread it on the surface of streams. It quickly dissolves and renders the fish senseless, so that they rise to the surface and are easily caught; but it does not harm them in any other way.

Early this century European and American farmers began to apply insecticides containing rotenone to various crops and found them very effective. The materials are used either as powdered root or as extracts obtained from the roots by solvents. Rotenone acts very quickly and has the great advantage of being harmless to man and other warm-blooded animals. It can therefore be used safely on food crops, or in insecticides, such as sheep-dips, that are applied to domestic animals. The barbasco and tuba plants are now cultivated widely. They are increased by stem cuttings, which grow into straggling bushes with long stems, and yield large roots two years after planting.

Another potent insecticide, *pyrethrum*, is obtained from the flower-heads of a small plant belonging to the *Chrysanthemum* genus. The plant is raised from seed on a large scale in America, Japan, and East Africa. Once the plants are established, they yield annual crops of daisy-like flowers, which are quickly dried in the sun or by artificial heat. The active ingredients, called pyrethrins, are usually extracted from the flowers by oil solvents. They are then made up into liquid sprays or powders of standard strengths. Pyrethrum insecticides are remarkable for their quick "knock-down" effect, which renders the insect helpless almost at once, though they do not always kill it; for this reason they are often mixed with stronger, but slower-acting, chemical insecticides.

Rotenone from the roots of the derris plant (above) is used against apple, maize, and bean pests and also against lice, fleas, and ticks on livestock.

Below: harvesting the daisy-like flowers of the pyrethrum plant. The main source of pyrethrins are the mature ovaries in the flower heads.

12 Oils, Resins, Rubber, Dyes, and Tanstuffs

Many plants manufacture substances in quite small amounts that man uses for a variety of purposes. We have looked at some of these in the two previous chapters. In this chapter we consider a few more. Some, such as oils, are edible and are used as aids to cooking; many others, however, are used as raw materials in industry, notably the chemical industry.

In seeking materials with which to make tools and other objects, primitive people naturally turn first to objects that are most familiar to them—pre-eminently to the plants and animals that provide them with food. The value of plants as providers of building materials —timber, rushes, and so on—was realized quite late in the total span of human history. Wool from sheep, hides and horn from cows, bones from these and other animals, were "fringe benefits," slowly discovered bonuses to add to the animals' primary role of food providers. As man's knowledge of nature has developed, and his material and cultural needs have grown more complex, he has learnt to look elsewhere

Right: a few of the most important oil-bearing plants. Percentages refer to the proportion of total oil content of each seed or kernel normally extracted commercially. Colour key: ochre—oils used mainly in foods and for cooking; ochre-grey—oils used both industrially and in foods; grey—non-edible, mainly industrial oils.

Oil-palm fruit

Olives

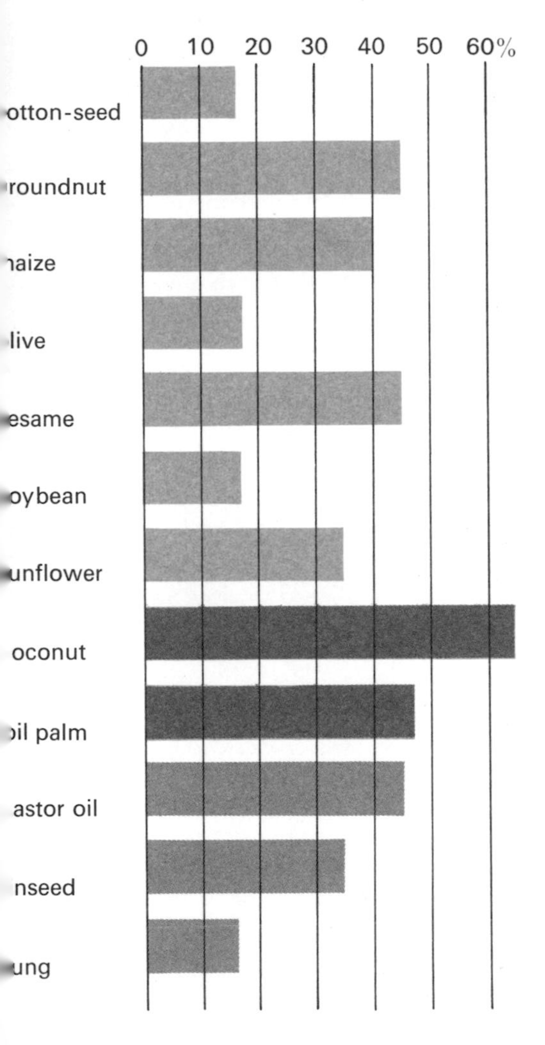

for his resources. He has, for instance, replaced timber fuels with coal and petroleum; and, as a more or less direct consequence, he has transformed his material life. Today, more than ever, man seeks new materials with which to make things. Much of modern technology is concerned with the search for synthetic materials to replace those formerly obtained directly from plant or animal sources. As we shall see in this and later chapters, many plant products are fighting a losing battle against alternative, man-made materials that are not only better for the job but can be produced more cheaply and in greater abundance.

Oils from Plants

Mineral oils, such as petroleum, have one great drawback: man cannot digest them, and so they cannot be used, as the plant oils and animal fats are, in foodstuffs. Mineral oils, however, have supplanted the other oils as fuels, as lubricants, and to a large extent in the manufacture of soaps and detergents. This development is very important for the nutrition of the world's growing population. For the rise of mineral oils has meant that plant oils, which were once used in lamps, and to lubricate machinery, and to make soap, are now available almost solely for use as food for hungry people. Oils—which may crudely be described as fats in liquid form—are a good, concentrated food, supplying ample energy from small weights of material; further, they are easy to store from one annual harvest to the next.

Cotton-seed

Peanuts (groundnuts)

All the commercially valuable plant oils are obtained from seeds or else from the pulpy coating of flesh that surrounds the seeds. Until recently, oils were obtained from a seed or pulp simply by crushing them in a press. Nowadays, other methods of extraction, such as heating, steaming, macerating (soaking), or applying solvents, help to raise the oil-yield. But some form of pressing or crushing is almost invariably needed as well because the oil is stored within the cells of the seed or pulp, and the cell walls must be mechanically broken before they will yield their harvest.

Oils may be one of two kinds: the drying kind, which evaporate when partially exposed to air and harden into a smooth, solid substance; and the non-drying kind, which do not evaporate and so remain in liquid form indefinitely.

The most important oil-yielding plant in the world, in both history and modern commerce, is the olive (*Olea europaea*). Truly wild olive-trees are hard to find, for every grove attracts groups of people who claim the crop and, to some extent, tend the trees. Apparently the tree once grew wild on the fringes of the Persian deserts, where it was brought into cultivation about 5000 B.C. The tending of the tree, and the use of the oil, were widely practised by the ancient Assyrians, Egyptians, Greeks, and Romans, and there are frequent references to olives in the Bible. In classical times the olive was carried westward to Greece, Italy, Spain, and North Africa. In more recent times it was exported by settlers to California, Mexico, South Africa, and Australia.

The olive-tree is adapted to the Mediterranean climate, with rainy winters and dry, hot summers. It is raised from cuttings and grows into an irregular, spreading tree that endures for centuries. The hard, strong wood is yellow in colour, with bold black or dark-brown markings, and as timber is valued for decorative carving. Olive leaves are evergreen and slender, with a dark-green upper surface and a silver-grey lower surface. After some eight years of growth the trees bear small, greenish-yellow flowers, followed during the summer by the slowly ripening olives, each on its own short stalk. In autumn they start to fall, and the olive-grower begins the tedious work of gathering. Cloths or mats are spread below the trees, and the fallen olives are picked up, one at a time, from the ground. The branches are beaten with poles to dislodge the late-ripening fruits.

Most of the crop is taken at once to the oil presses. At one time these were huge stone vessels, in which a

Above: olive orchard in south-western Spain. Semi-wild groves of olive-trees have been tended by Mediterranean peoples since the days of the ancient Egyptians.

round stone was turned by hand power, to crush the olives and force out the oil. Nowadays crushing by power-driven grindstones is followed by pressure in a hydraulic press, and then by solvent extraction of the residues. The pulp is fed to cattle.

Olive-oil is the main cooking fat of the Mediterranean lands, where it takes the place of the butter or lard used farther north. It is very nutritious, and fetches a high price. Oil of low grade—from damaged olives, for example—is used for making soap, as a lubricant, or as a fuel in oil lamps.

Raw olives have a bitter flavour and cannot be eaten when they first fall from the tree. To improve their flavour they are steeped in brine for many days. Some are removed from the brine while they are still green; others are pickled for a longer time, so that they become almost black and acquire a stronger taste.

Other important oils derived from tree crops include coco-nut oil and palm-oil (described in Chapter 9). Another is tung-oil, which is obtained from the seeds of the tung-tree (*Aleurites fordii*), a native of China, and is a powerful drying agent used in paint and varnish.

Below: olive-harvesting methods (left, from a classical Greek vase; right, on a Spanish plantation) have remained much the same for more than 2000 years.

Olive-oil is widely used as a cooking fat. Green olives, as above, are pickled for several months in a salt solution; black olives are also pickled but are allowed to ferment.

Oils from Annual Field Crops

The olive-tree does not thrive in the cold, damp climate of northern Europe, so at an early date the northern peoples sought a local substitute. They found this in the seed of a small cabbage-like plant called rape or colza (*Brassica campestris oleifera*), which is now grown in Europe and America as an annual crop like grain. Its small, round, dark-brown seeds are threshed from their pods, and then pressed to force out the oil. The seed-cake that remains is fed to pigs or cattle.

Oil from the seeds of rape (grown in China, as here, and India, as well as in Europe) is a useful alternative to olive-oil; it is also used as a fuel and as a lubricant.

Many other oily seeds have been used commercially at different times and in different lands. A major crop in Russia is the beautiful annual sunflower (*Helianthus annuus*), which came originally from Mexico. Sunflowers are grown from seed, like a grain crop, and given a fertile soil and hot summer weather they produce a stiff, upright stem, six feet tall, bearing many large, simple leaves. At the tip of this great stem the plant bears one huge, round flower, as much as a foot in diameter. Like all the flowers of the daisy family (Compositae), to which the sunflower belongs, it is a compound flower consisting of many florets. Those in the centre, which make up the "disc" of the flower, are small and dark-yellow in colour; those around the rim, called the "ray florets," have bright-yellow petals that stream out like the rays of the sun. The huge flower is borne sideways at the top of the stem, and it always faces toward the sun. In this way it attracts the insects that pollinate its florets.

Each floret ripens a single seed, which is large and oval in outline, but rather flat, and may be white, black, brown, or striped. The crop is harvested by

Below: sunflower-seed oil is used as a cooking oil and in the manufacture of soaps, lubricants, and paints. As is the case with many other edible oil-seeds, the cake remaining after the oil has been extracted is highly nutritious and is used as a cattle feed.

cutting off the soft, pithy, fruiting heads, and scraping the seeds out of them.

The cotton plant (described in Chapter 13) provides another important oil, though the plant is never cultivated for the oil alone. Before cotton fibres can be spun into thread, all the little seeds, to which they are attached, have to be removed by a machine called a *gin*. At one time the seeds were thrown away; but nowadays they are pressed to yield cotton-seed oil, which resembles olive-oil.

Most grain crops hold little oil in their seeds. But maize is an exception. The oil in the maize grain is not easy to extract, but it is valuable because it has much the same properties as olive-oil, and can be used for eating and cooking. Some maize farmers in the United States and elsewhere grow their entire crop for "corn-oil," as it is called. It can be a profitable business, because the residue from the oil-extraction process is rich in carbohydrates and proteins and can be sold as a stock feed.

Linseed-oil an edible drying oil, is extracted from the seeds of flax plants (above). A native of central Asia, flax is cultivated throughout the temperate zones, notably in Argentina.

Linseed-oil is a good example of the drying oils that form the base, or medium, of many paints. It is liquid when applied, but dries later to form a firm, smooth, waterproof surface that is useful for protecting wood and other materials against moisture. Linseed-oil is also used to make linoleum floor-coverings, which are compounded with cork and have a jute fabric as their base. The oil is obtained from the small, flat, brown seeds of the blue-flowered flax plant (*Linum usitatissimum*), which is also widely cultivated for its fibre, from which linen is made. As a rule the farmer who is concentrating on the oil crop uses varieties that have been bred to yield much seed but little fibre.

Below: gathering the heads of black sunflowers on a farm in Kenya. The plants commonly reach a height of between 12 and 14 feet.

Wax from a Palm Tree

Waxes are chemical compounds called *esters*, and consist of hydrocarbons in combination with certain forms of alcohol. The best-known example is bees-wax, which bees secrete in order to make the honeycomb. Unlike oils, waxes are usually solid at atmospheric temperature, though they melt readily if heated. This means that they can be used for polishes, for example on furniture, to give a hard, smooth, and attractive finish. They are highly repellent to water, so they are used for waterproofing textiles, wood, and leather. The most important domestic use of waxes is in shoe polishes.

Many plants produce natural waxes; in fact, no plant can live in a hot, dry climate without some form of waxy substance on its leaves to lessen the loss of water due to evaporation. But on most plants the quantities are small, and not worth collecting. The main commercial source of plant wax is the carnauba palm tree (*Copernicia cerifera*), which grows wild on the hot, dry savannahs of north-eastern Brazil. It is almost the only tree to survive the semi-desert conditions, and it grows only in open groves, with broad gaps between the trees. The wax forms a coating on the surface of its fan-shaped leaves, and a proportion of these are cut from each tree during the dry season, when the wax content is highest; if too many leaves were removed, the tree would die. The leaves are taken to drying sheds, where they are cut into shreds, by hand or machine, and allowed to wither. Then they are beaten, so that the wax falls off. The wax is then melted and moulded into blocks for export. After refining, it is used in the manufacture of shoe-, floor-, and furniture-polish, gramophone records, carbon paper, matches, crayons, and plastics.

Carnauba palms of coastal Brazil yield high-quality wax for shoe, furniture, and floor polishes. The wax forms as a coating on the leaves to prevent loss of moisture during the hot, dry summer. The leaves are cut, in September and December, with knives on the end of long poles (above), and dried (below, left) in sheds for about two weeks. Then they pass into a tumbler, where the dry wax is knocked off and, now in powder form (below, right), is bagged for export.

Resins from Pine Trees

Most cone-bearing trees of the natural order Coniferae produce chemical compounds called *resins*, which usually occur in their leaves, buds, and cones, as well as in their wood. These resins flow through fine channels called resin-canals, and their purpose seems to be a protective one. The resin consists of two elements, *turpentine* (a volatile substance) and *rosin* (a solid material). If some of the bark of coniferous trees is cut away, the resin oozes out, the turpentine evaporates, and the rosin is left as a solid layer over the wound. If an insect, such as a bark beetle, attacks a healthy tree, it may be drowned in the resin that flows out as soon as it bites it. The solid rosin that forms on damaged tissues stops the spores of many harmful fungi from germinating effectively on broken branches and other parts of a tree that have had parts of their bark destroyed.

The highest concentrations of resin are found in the trunks and roots of certain pine trees that grow in hot, dry countries. Forests of maritime pine (*Pinus pinaster*) are valued for resin production in south-western France, and forests of long-leaf pine (*Pinus palustris*) and related species are managed for the same object in various areas of the United States. In both countries, the pines are raised primarily for timber, but resin is a valuable by-product. The resin flow is seasonal, the trees being tapped in the spring.

To win the resin from the pine tree, it is necessary to cut grooves deep in the trunk through the bark and bast layers and into the living wood. This can be done on only one side of the tree at a time, for if the cut extended all round the trunk the circulation of sap would be stopped, and the tree would die. The resin tapper cuts the groove with an extremely sharp, curved chisel, or gouge. Resin oozes into the groove, and is led down another, vertical groove into a spout, and so into a cup. In order to keep it flowing, sulphuric acid is applied to the resin; this prevents it solidifying for about three weeks. After this time the tapper returns and freshens the cut by extending it upward and applying more acid; he also collects the resin from the cup. Gradually the cut extends upward in a panel on the tree, which will yield a seasonal harvest for many years before the whole tree is felled for timber. One tapper can deal with 10,000 trees, each yielding a few ounces of resin a year.

The resin is taken to a distillery, where hot steam is used to separate it into its two component substances. The turpentine comes away as a light spirit, and the rosin is left behind as a solid, yellow substance.

Tapping long-leaf pine for resin in a forest near Stockton, California. The long-leaf and its close relative slash pine, both native to America, supply the bulk of the world's supply of resin. The two by-products of resin—turpentine and rosin—are used mainly in paints and varnishes; rosin is also used in printing inks and for coating certain types of paper for printing. Both have been partially superseded by synthetic materials made from mineral oils.

Tapping gum arabic (above) in Kordofan, Sudan. The small, gnarled trees exude gum after their outer bark has been stripped. The gum (below) solidifies in a few days and is then cleaned (bottom) and bleached in the sun.

Gum from a Desert Tree

In the deserts of north-eastern Africa, Arabia, and north-western India grows the gum-arabic tree (*Acacia senegal*), a member of the mimosa family. It has a short trunk and spreading branches studded with hooked thorns that protect it from browsing camels and antelopes. It is leafless in winter, but in the summer it bears feathery, compound leaves and pretty yellow mimosa-like flowers, followed by pea-like seed pods. Each spring, after the rains, nomad tribesmen such as the Somalis (who claim traditional rights in groves of gum-trees) tap the trees. They make axe cuts in the branches and strip off patches of bark. The clear, pale-yellow gum, which is the tree's natural defence against wounding, oozes out and solidifies. The tappers return a few weeks later, scrape it away, and dry it in the sun. It is then despatched by camel train to market. Gum arabic, which is tasteless, odourless, and slightly soluble in water, is used as an adhesive, and also as an ingredient of inks, polishes, and size, as a "glaze" in confectionery, and in medicines.

Sugar from the Maple Tree

The sap of all trees is used for the transport of dissolved sugars, for this is the only way in which the carbon compounds needed for energy and for the building up of plant tissues can be moved from one part of the tree to another. The main flow of dissolved sugars is downward, in the channels of the bast tissue just beneath the bark, because the food is manufactured in the leaves. But during a few weeks in the spring there is, in some trees, a strong upward flow of sugar-rich sap within the wood itself. This happens because such trees store food reserves in their lower trunk and roots and need them high in their branches to nourish new leaves and flowers.

Birch trees and maples have a strong flow of sugar-rich sap in their sapwood at this time, and if the wood is injured, they "bleed" freely. But only the sugar maple (*Acer saccharum*) of the north-eastern United States and eastern Canada yields enough sugar to be worth tapping commercially. Farmers who own maple groves get a valuable harvest at the time of the spring thaw, usually in March. They drill a hole deep into the wood of each tree near the base, insert a metal spout, and set a small bucket below this to catch the sap flow, which lasts for a month. Alternatively, plastic tubes are led from several trees to a single bucket. The sap must be harvested daily or it will ferment and turn sour. It consists of 97 per cent water and only 3 per cent sugar. It is concentrated by evapor-

French engraving of 1724 (above) shows American Indians gathering maple sap and converting it into sugar by boiling.

ation in open iron pans over wood fires, to form a deep-brown syrup containing 66 per cent sugar; in this form the sugar will keep well. Further concentration causes the sugar to crystallize out as a pale-brown solid called *maple candy*.

Maple syrup was known to the Indians, but they found it difficult to concentrate, as they had no metal pans. The early colonists of New England relied on it, rather than on cane sugar, but today it is considered a delicacy—relished for its nutty flavour throughout Canada and the United States—but unimportant in the world sugar market.

Below: collecting maple sap in Canada before the spring thaw (left). The sap is tapped by drilling one or more holes, about two inches deep, into the trunk of the tree, inserting a spout and collecting the fluid in a bucket (right).

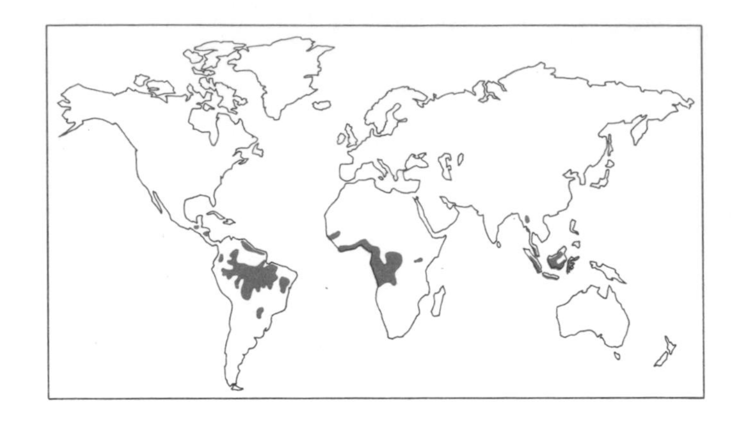

Map (left) shows main rubber-producing areas of the world. The principal sources are Malaysia, Indonesia, Thailand, Ceylon, South Vietnam, and Nigeria. The current world production is about 2.3 million tons a year.

Below: the rubber tree bears palmate leaves and a three-lobed seed capsule. The mature trees vary in size; in the wild, 130-footers are not uncommon.

Rubber

Nearly all the natural rubber used in the world today comes from the Pará rubber tree (*Hevea brasiliensis*). There are other plants that yield rubber-like substances, and there are also synthetic rubbers that are better for some purposes; but it is Pará rubber, sometimes pure and sometimes compounded with other substances, that is used where the need is for something elastic, hard-wearing, and impervious to water or air. Tyres for cars and lorries are a key use, but rubber is also widely used in flooring, sheeting, and in many industrial applications where the natural product is better than any synthetic alternative.

The Pará rubber tree grows wild in the jungles of northern Brazil and is named after the port at the mouth of the river Amazon that was once the world centre of the rubber trade. It is a tree of medium size, with a smooth grey bark, compound leaves (each with five leaflets), small greenish-yellow flowers, and large smooth seeds. It can stand dense shade and grows

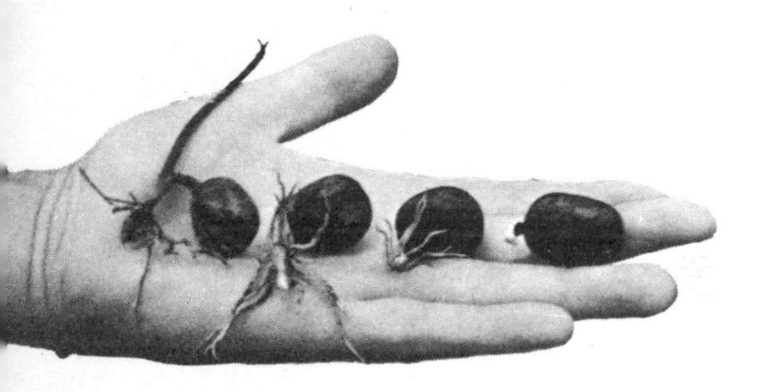

Seeds germinated in a specially prepared seed-bed (above) are ready for planting in this Malaysian nursery (right), each seed in its own basket. These seedlings, about six weeks old, are almost ready for transplanting in the field.

amongst many other, taller species in a region of constant heat and high rainfall. In all its surface tissues, though not in its wood, it has special channels that contain a white, milky fluid called *latex*. Whenever a rubber tree is damaged, latex flows out and, within a few hours, solidifies and seals the cut with a natural coating of pure rubber.

Rubber was at first a curiosity, used by the Amazonian Indians for making balls. But when, in 1839, the American inventor Charles Goodyear discovered how to make it more durable by *vulcanization*—heating it with sulphur—the demand for it rose steadily.

The rubber tapper (below) shaves off a thin strip of bark half way around the trunk. Latex oozes out of the bast into the cup suspended below the strip. At each tapping the latex flows for four hours or so—enough to fill a teacup.

In 1876 the British Government of India commissioned an Englishman, Sir Henry Wickham, to bring seeds from Brazil. He did so not by smuggling them out (as a popular story goes) but with the assistance of the Brazilian government. A few weeks later the first rubber trees to be grown outside Brazil sprouted in the greenhouses of the Royal Botanic Gardens at Kew, near London. Only a few months after this, the little seedling trees were sent out in glass cases to Ceylon, to become the foundation stock of the vast rubber plantations in India, Ceylon, Burma, Malaysia, and Indonesia that meet most of the world's demand. The climate and soil of south-eastern Asia proved ideal for the tree, and Indians, Indonesians, and Chinese quickly learnt how to tend it and harvest its latex. Nearly half the world's rubber comes from small holdings, the rest from big plantations run by European or American companies. Plantations have also been established in Liberia and Brazil.

Under cultivation, the rubber plant is always raised from seed. Unless the seed itself comes from a high-yielding strain, the plant is later grafted with a bud from a high-yielding variety. Its main stem is then cut back almost to ground level, and it grows—from the bud—a new trunk that will give ample latex later on.

Rubber trees are planted, about 15 feet apart, on cleared jungle land that is kept free of re-invading wild plants by growing a cover crop; this also checks soil erosion. The trees grow fast and can be tapped only five years after planting. Each tree is tapped, on

Above: transplanted seedlings, partly obscured by a soil-enriching leguminous cover crop, arise amid old, poisoned rubber trees on a Malaysian plantation.

Below: at nine months the seedlings are grafted with a "bud-patch"—a strip of wood taken from a high-yielding tree. After a few weeks a shoot grows from the graft and the stem of the original plant is cut off a few inches above it.

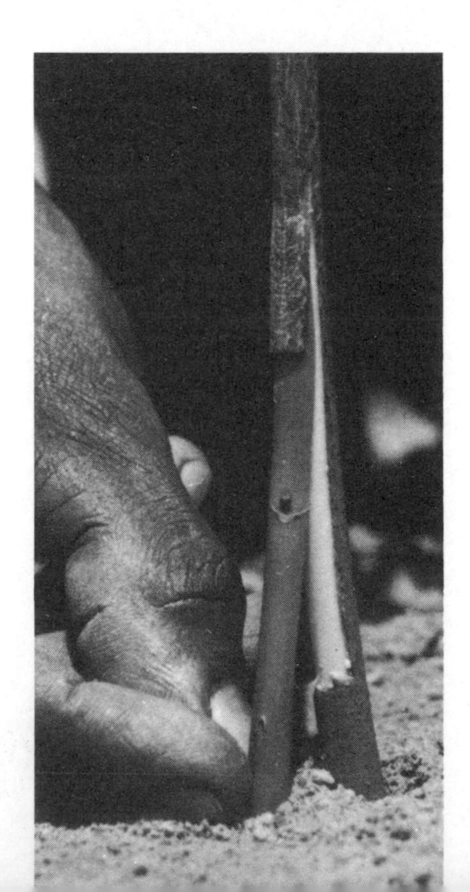

average, every third day by tappers who visit some 400 trees each day. Over the course of the year, each tree will yield two pounds of dry rubber. Tapping involves cutting a groove halfway around the tree, sloping downward spirally to a vertical channel also cut in the bark. The white latex flows out of this groove, and so down into a little metal spout, which carries it to a cup fixed on a wire. At each fresh tapping the groove is re-opened, working downward, so that a broad "face" is gradually formed on one side of the tree. If tapping is skilfully done, the bark renews itself and can be tapped again a few years later; in the meantime, the other side of the tree is tapped.

After tapping in the cool of the morning, the tapper returns to the trees with a bucket, and pours into this the latex that has flowed into each little cup. He may carry to the factory about 60 pounds of latex, and of this one third is solid rubber, the rest water. Several methods are used nowadays for concentrating the latex, or solidifying it as rubber, for shipment. The simplest one, called the sheet process, is to pour it into a big aluminium tank divided lengthwise into narrow sections, and to add a few drops of formic acid, which makes the latex coagulate into a spongy mass. This is run between series of rollers, which squeeze out the water, so forming sheets of pure white rubber. The wet sheets are dried over a smoky wood fire that coats them with creosote and so stops moulds forming; then the smoked sheets, which are now amber-coloured and translucent, are packed for export.

Latex is fed into a centrifuge (above). From the spout at left concentrated latex emerges; from the lower right-hand spout, a mixture of water and rubber particles, from which a crepe rubber is made.

Below: acid added to the latex in the tank (left) makes the latex coagulate into sheet rubber. The sheets pass through rollers and are cured for about 48 hours in a smoke-house (right).

15th-century French manuscript shows dyers stirring a vat of woad dye to ensure uniformity of colour. The red cloth on the side of the vat is about to be dipped; the cloth on the floor is perhaps being used as a matching sample. Mediaeval dyers fixed woad dyes with alum or potash mordants. Like indigo, woad assumes its final colour only after it has been dried and exposed to oxygen in the air.

Below: four egg shells coloured with readily available plant dyes. From left: two different concentrations of onion-skin dye; black currant; turmeric.

Dyestuffs

Until about 150 years ago most fabrics were coloured with dyes obtained from plant, animal, or mineral sources. But early in the 19th century chemists began to experiment with dyes made from various chemical compounds, notably *aniline*, which is extracted from coal tar. In the intervening years they have been so successful in developing substances that give lasting, vivid colours at low cost that plant dyes are now of only historical interest in the commercial world, though they are still used by craftsmen who work on a small scale or in remote places.

Most of the plant dyes come from wild-growing plants, for the quantities needed can usually be gathered without cultivating a crop. The dyes are found in various parts of plants, from wood to leaves or fruits. The colours they yield depend much on the dyer's skill, and on the chemical *mordants*, such as alum, that fix (render insoluble) the dyes when they are applied to textiles, wood, or other materials. Onion skins give orange colours, the roots of the madder plant (*Rubia tinctorum*), produce bright red, and the heartwood of the Siamese jack-fruit tree produces the brilliant yellow colour seen on the robes of Buddhist monks. In northern countries, lichens are often gathered as sources of dyes; a yellow lichen found on the shore of the Scottish islands gives a rose-pink hue.

A few plants were formerly cultivated to yield dyes needed in quantity. Most important was indigo (genus *Indigofera*), believed to have originated in Indonesia. It belongs to the pea family and bears compound leaves and spikes of tiny reddish-yellow flowers. The dye is made from the leaves, which when left to ferment in water produce a paste that is dried to form the marketable product. This is colourless, and the deep blue shade appears only after the dyed fabric is exposed to the air. Woad (*Isatis tinctoria*—a member of the mustard family) was once widely grown in Europe; at the time of the Roman conquest it was used by British warriors as a war-paint. The plant is raised from seed and bears bright-yellow flowers. Woad leaves have to be steeped in water before they yield the paste which, after drying, re-wetting, and further fermentation, gives a blue dye.

Madder dye with alum mordant

Madder dye with chrome mordant

Madder dye with tin mordant

Heather dye with tin mordant

Weld (wild mignonette) dye with chrome mordant

Ragwort dye with chrome mordant

Lichen (*Parmelia physodes*) dye; no mordant

Walnut-bark dye with iron mordant

Yew-chips dye with chrome mordant

Indigo vat dye; no mordant

Wool samples at right have been dyed with dyes named in captions above; the plant sources are shown in sketches. Note effects of different mordants on the madder dyes.

Tanstuffs

The art of tanning hides and skins to turn them into curable leather has been known from very early times. If the skin of an animal is left untreated, it soon decays because bacteria start to feed and grow on it. But certain plant compounds, called *tannins*, precipitate the proteins in hides and stop the growth of bacteria, so producing a material that is both strong and supple, and will remain so for many years. Tannins are most abundant in the inner bark of certain trees. They are easily obtained from the bark by soaking it in water, to form a tanning solution in which the hides are left to soak—often for many weeks.

Oak bark is particularly rich in tannins, and for centuries it has been stripped for this purpose from standing trees about to be felled, or from freshly felled logs. It "runs," or comes away, most freely in the spring. The bark is then dried with its outer side upward, so that the soluble tannins do not get washed out by rain. Other trees used for tan-bark are spruce (genus *Picea*) in Europe and hemlock (genus *Tsuga*) and chestnut (*Castanea dentata*) in North America. These barks are by-products, harvested when the trees are ripe for the timber for which they are primarily grown; but they bring in a good return to the forester.

Wattle bark, which is now the most important kind grown commercially, comes from an Australian tree (genus *Acacia*) that bears feathery leaves and pretty, yellow blossoms. The main plantations are in South Africa and South America. The trees are grown from

Oak bark, seen above being stripped from a log, is the principal source of tannins in Britain. After removal from the tree, Norway-spruce bark (left, from Schneegarten forest, Austria) is dried on racks before its tannins are extracted. With future supplies of tan-bark uncertain in the temperate zone, tanners are looking elsewhere for plant sources. These include the bark of the mangrove, a tropical tree, and the roots of canaigre, a farm crop that somewhat resembles the parsnip. Synthetic tannins are also being developed.

Above: hides are dipped in vats at the tanning works. Hides are usually prepared for tanning by immersion in a solution of lime to loosen the hair, which is then scraped off. The hides pass through tanks containing progressively stronger tanning solutions, so that the innermost fibres are thoroughly tanned.

Below: tan-pits in Marrakesh, Morocco. An extremely ancient craft, tanning was practised by the Egyptians and Chinese at least 3000 years ago.

seed and harvested 10 years later when they have formed tall poles. The bark, which contains 50 per cent tannin, is stripped off with sharp knives, and the bare poles are sold for pulpwood or pit-props. Some of the bark is powdered for direct use, but in most cases the tannin is extracted and marketed as a liquid extract. The demand for plant tannins has declined in recent years as chemists have synthesized a variety of tanning materials from other sources. But tannins continue to be used, on a small scale, in such diverse applications as preventing the formation of scale on the inside of boilers, and clarifying wines and beers.

13 Fibres, Roofing Materials, and Cork

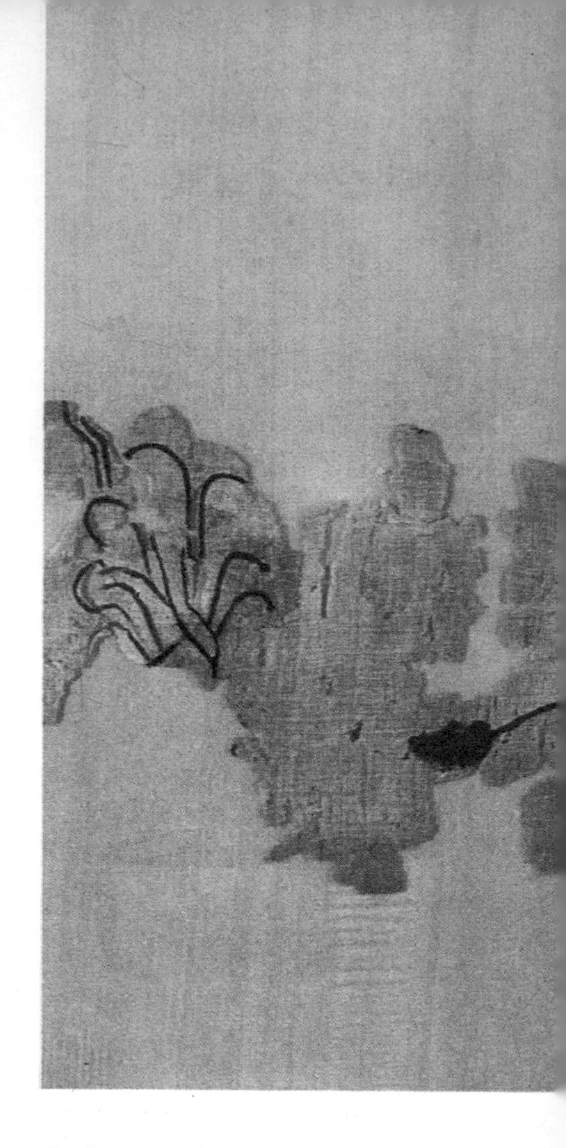

The fragment of linen cloth (above), from Egypt, dates from pre-dynastic times. The craft of weaving enabled prehistoric man to supplement hide and fur clothing.

Man's use of fibres began deep in prehistoric time. Probably his first attempts to make ropes for trapping and hunting consisted of twisting bundles of wild grasses together. Much later, he discovered not only that certain plants contain very fine fibres but also that, with skill and patience, he could separate the fibres out from the rest of the plant, twist them together to form a yarn, and so make fishing nets and crude textiles. He also learnt to use the wool and gut of animals—sheep, goats, llamas, even dogs—but once he had discovered the arts of twisting, weaving, and tying knots, he found that plants gave him a much larger, more readily available supply of fibres that could be put to a variety of uses.

Fibres occur in all but the simplest plants. They consist of specialized cells that give pliant strength to the plant tissues. When a wheat stem faces a strong wind, it is the fibres that enable it not only to bend without snapping but also to spring back again when the wind slackens. It is the fibres too that enable the stem of a great tree to remain flexible while supporting the enormous weight of its branches.

There are, however, few plants that contain fibres of economic importance to man. Most wood fibres, for instance, can be extracted only by breaking down timber with powerful machinery, and even then the fibres are usually too short to be twisted into yarn (though wood pulp is the basis of rayon). Fibres in the stems of wheat and many other plants are useless because they become brittle when dry. The plant fibres we use today are of one of three kinds: seed hairs (as are found in cotton and kapok); stem fibres (flax, jute, and hemp); and leaf fibres (sisal and others). Seed hairs require little processing before they are made into yarns, but the other types of fibre must be separated out from their enclosing tissues. A seed hair is also structurally different from a stem or leaf fibre. A cotton fibre, for instance, is a *single* cell, about 3000 times as long as it is thick; one pound of cotton may contain 90 million seed hairs. Stem and leaf fibres,

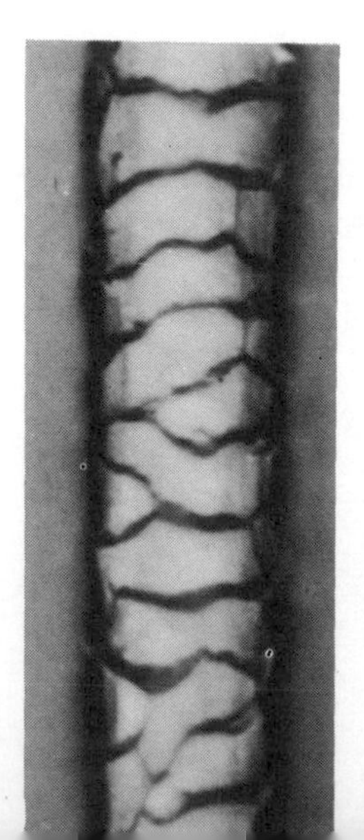

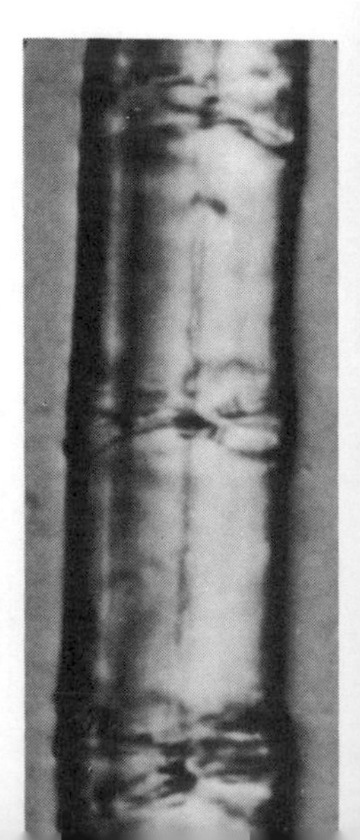

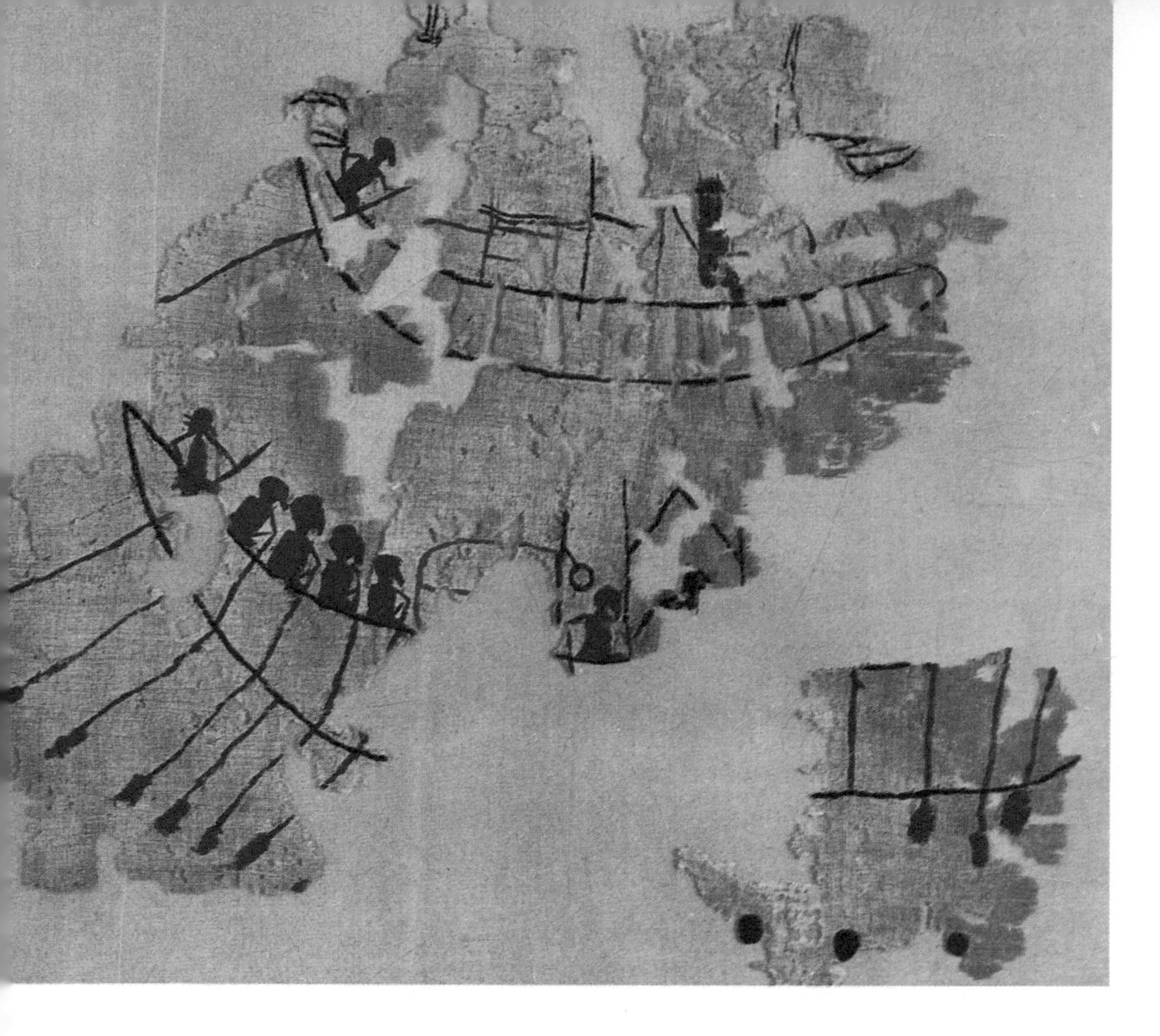

The fibres (below) are each magnified 500 times. From left: wool, flax, cotton, and rayon. Note the twist in the cotton, which occurs in the making of yarn.

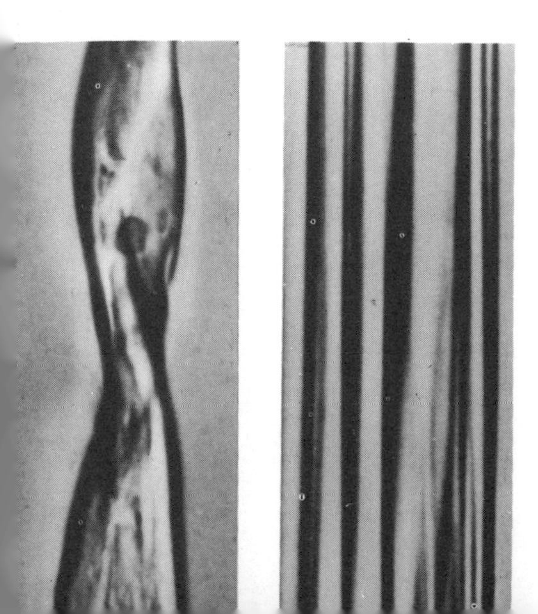

on the other hand, are made up of *many* cells growing closely together as part of the vascular bundles of conductive tissue; although such fibres may be very strong, they yield a coarser yarn than cotton.

During the last 75 years or so, plant fibres have been replaced in some uses by synthetic fibres—the products of the chemist's laboratory. The first of these synthetics was rayon, developed during the 1890s. The best known, of course, is nylon, which was first marketed (in the form of stockings) in 1940. The main advantages of synthetic fibres are strength (nylon is twice as strong as cotton) and resistance to moth and bacterial attack. Both natural and synthetic fibres have an important role to play, however, especially in clothing textiles, where the material often comprises a mixture of fibres in order to exploit the advantages of both types; cotton, for instance, is more comfortable next to the skin because it absorbs perspiration, while synthetics are harder-wearing and easier to wash.

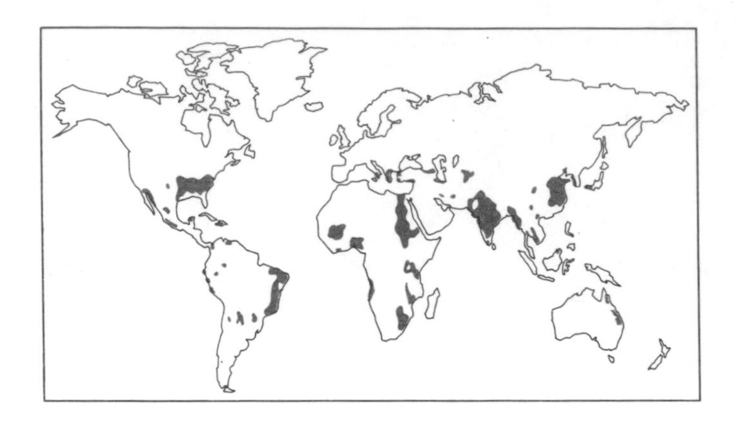

Map (left) shows main cotton-producing areas of the world. Cotton-plant (below) is ready for harvesting: the fluffy white fibres have emerged from the bolls, which have opened in the autumn sun.

Cotton

Cotton is still the world's leading fibre for spinning thread and weaving cloth. It is made from the seed hairs of annual plants of the genus *Gossypium* that have been bred from wild species native to subtropical Asia and America.

In prehistoric times people found that when they twisted the cotton hairs together they did not spring apart but remained as a firm, strong thread, or yarn. When a number of these threads were interwoven in criss-cross fashion, they held together and formed a tough, elastic cloth that, while thin and tight, was hard-wearing and could be washed and dried repeatedly; moreover, it could easily be dyed or printed with colours. The spinning and weaving of cotton cloth became a cottage industry in India, Indonesia, and other eastern countries, and also among the Aztecs of Mexico. But it was slow and tedious work, for the fibres are very fine and only half an inch to two inches long.

The cotton plant, and the arts of cotton spinning and weaving, were introduced to Europe by the Arabs in the 7th century, and Sicily and Spain, as well as North Africa, became centres of cotton production. Spanish settlers took the plant to the West Indies in the early 16th century, only to find that the Indians already grew and used a native species. It was found to thrive also in the southern states of America, where great plantations were established to meet the needs of Europe, and thousands of African slaves were shipped over to tend this new and profitable crop.

During the 18th century the wool spinners and weavers of the county of Lancashire, in north-western England, also began to use cotton. The moist, rainy climate helped them to handle the thread without breaking it (cotton gains in strength when wet), and plantation supplies had made the raw fibre cheaper

than raw wool. Toward the end of the 18th century a group of inventors, James Hargreaves, Samuel Crompton, and Richard Arkwright, developed machines that enabled them to spin cotton thread much faster than by hand. Another breakthrough came in 1785 when a Nottinghamshire parson, Edmund Cartwright, invented a steam-powered loom. The first reaction of the cotton workers was to destroy these machines, for fear they would lose their jobs. But the machines were so efficient that they soon won the day, made cotton cloth a cheap fabric for everyday use, and, incidentally, precipitated the Industrial Revolution.

For a hundred years the Lancashire spinners and weavers dominated the world trade; they even sold cotton cloth to India more cheaply than it could be made there by hand methods. But eventually other countries set up their own mills, and thriving cotton industries are now established in most parts of the world.

Though cotton is used everywhere, it is grown in comparatively few countries. The United States, Egypt, Sudan, the Soviet Union, Brazil, and India are the main suppliers, while substantial amounts are exported from China, Nigeria, South Africa, and the West Indies—the last named being famous for its long-fibred sea-island cotton. The plant thrives best in a subtropical climate with spring rains and a hot, dry summer. The cotton plant is an annual, some three feet high, and is raised from seed sown direct on well-tilled land, in rows about three feet apart. It needs

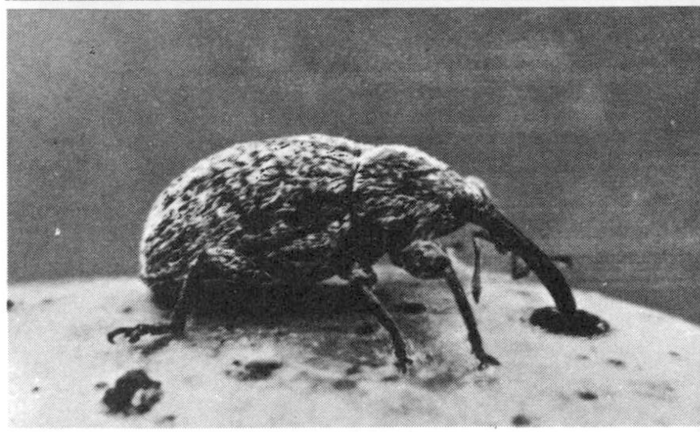

Above: cotton-plants (upper picture) are sprayed with insecticides to protect them against the boll weevil (lower picture), seen here boring into a boll.

Below: the cotton harvester straddles each row of plants, the cotton being pulled from the bolls by barbed spindles revolving on vertical drums.

Left: the harvest is fed into a cotton gin, which separates fibres from seeds and removes other unwanted material such as stalks and leaf fragments.

careful weeding, and in dry countries the fields are often irrigated. After eight weeks of growth the plants produce many large, bell-shaped flowers in shades of bright-yellow, orange, red, or white, and at this time the fields are a magnificent sight. Each flower is succeeded by a seed pod, or *boll*, which is brown in colour and holds many small, black seeds. These seeds bear slender white hairs—the cotton fibres—whose role in nature is to support the seed on the wind and aid its dispersal. As the bolls ripen and turn brown, they split to release the seeds. The crop is gathered in stages, because the bolls ripen at different rates. Formerly, picking was all done by hand, and a big force of cotton-pickers—men, women, and children—was needed at harvest time. Today boll-picking machines deal with much of the crop; but to enable them to operate, the plants must first be stripped of their leaves by applying a poisonous spray such as calcium cyanamide. One acre of cotton yields, on average, 400 pounds of fibre and 800 pounds of seed. It is an exacting crop, needing ample fertilizers to maintain good yields and prevent soil exhaustion.

The bolls are taken to sheds, where the husks are removed and the remaining cotton passed through a *ginning* machine. This removes the seeds, which are used to make cotton-seed oil and oilcake. The cotton gin was invented about 1790 by an American engineer, Eli Whitney; by increasing the amount of cotton that one man could clean in a day from five pounds to a thousand pounds, it played a major part in cheapening cotton production. The gin works by drawing the cotton hairs through narrow slits too fine to allow the seeds to pass.

Cotton is used for many kinds of cloth, for sewing thread, and in industrial textiles, such as the strong fabric of motor-car tyres. Cotton-wool, a by-product, is used as a cleansing material and a wound dressing.

Below: the carding machine cleans and separates the individual cotton fibres. The thin web of fibres is then formed into a rope-like strand called a sliver.

Slivers are twisted slightly to make them thinner, then pass to this spinning frame where they are stretched, twisted again, and wound onto bobbins.

Kapok

Kapok fibre is the seed hair of a tree (*Ceiba pentandra*)—sometimes called the silk-cotton tree—that grows wild in tropical Asia and Africa, and is cultivated on small farms in Indonesia and the Philippines. During the early years of its life *C. pentandra* resembles a child's stylized representation of a tree: its trunk grows straight upright and its branches shoot out at right angles. It is raised from seed and after some seven years' growth it bears greenish-white flowers followed by many large seed pods. In the dry season, when the pods ripen, the tree is leafless; the pods are harvested with knives on long poles. They must be picked before they open, or else the fibre, called *floss*, will blow away. They are spread out on sunny drying platforms under a screen of fine wire netting. The sun dries the pods, so that they open, and the netting checks the wind and prevents the floss blowing away. The seed is separated by simple machinery, and the oil within it is extracted for use in soapmaking.

Though kapok fibre does not make a good thread, it is very valuable for two reasons. First, it is a poor conductor of heat and so makes a good "filler" for mattresses, and for sleeping bags, anoraks, and other garments worn as a protection against wind and cold. Second, it is quite impervious to water, and so makes an excellent floating or buoyancy material when used in sailors' life-jackets.

Fruits of the silk-cotton tree consist of large pods, the inner walls of which are lined with kapok floss. When ripe for picking, as here, the pods become wrinkled and the floss shows through their cracked surface. Unless harvested quickly, the cracks will open and the floss will be blown away by the wind.

The silk-cotton tree (left) is native to East Africa (as here), West Africa, India, Ceylon, the Philippines, and Indonesia. Under favourable conditions it attains a height of 100 feet and bears a full harvest in its eighth year.

Flax

Before cotton became a cheap and popular fabric, the only textiles available to the northern peoples for clothing were wool from the sheep's back and linen from the flax plant (*Linum usitatissimum*). The cultivation and use of flax are extremely ancient crafts. Traces of the fibre have been found in the prehistoric lake villages of Switzerland, and it was used by the ancient Egyptians to make wrappings for their mummies as early as 4000 B.C. Several plants similar to flax grow wild in Europe, and the cultivation of the common kind in small plots, sufficient for the needs of a single village, was a general practice in the Middle Ages. Much labour was needed to prepare the thread, as well as to spin and weave it, yet most communities made enough for their own needs. Today flax is grown as a field crop, mainly in Canada, the United States, the Soviet Union, India, and Brazil but also to a lesser extent in western European countries, such as Holland, Belgium, and France. Northern Ireland, where a little flax is still grown, is the main importing and manufacturing country.

Flax is grown as an annual crop, preferably in regions with a cool, moist summer followed by a dry autumn. It is grown from a small, flat, brown, oily seed (the linseed). As we have seen, the linseed raised for oil is simply another variety of the flax plant, with more branchy stems. Flax is raised rather like a grain

Flax-plant has blue or white flowers. Most subspecies cultivated for fibres have long stems, white flowers, small seeds, and are of Dutch origin.

Traditional method of harvesting flax (above, left, in ancient Egypt; above, in Northern Ireland). Machines capable of uprooting the crop are also used.

Above: flax is removed from a retting dam after soaking for a few days.

Below: combed tresses of flax—each tress overlapping, to make a continuous length of fibre—are laid on a spreading board. This carries them over sets of pins and between rollers so as to form them into lengths of yarn.

crop. Seed is sown densely to discourage the formation of tillers and to limit flowering and seeding. Those flowers that do open are bell-shaped and white. The growers aim to produce tall, slender main stems up to four feet high. To get the maximum length of undamaged fibre, the crop is harvested, in autumn, by pulling it out by the roots. This was formerly done by hand, but a tractor-drawn machine is often used today in North America and Europe.

The harvested flax is first dried, by stacking it in stooks or upright heaps, and then *retted* by wetting it thoroughly and then allowing the soft tissues of the stem to rot partially. The stems are thoroughly dried and then *scutched*—that is, beaten to free the softer tissues from the fibres, which form the *bast*, or outer layer, of the stem. Scutching was once done by hand beating, but nowadays it is achieved by passing the stems between rotating rollers and drums.

The fibres are pale yellow in colour and remarkably long—from one to three feet. They are uneven in width, for each is really a bundle of narrow fibrous cells, about one inch long. They are flexible, yet very strong, and (like cotton) become even stronger when wet. Flax is used wherever a strong thread or fabric is needed, notably for sewing thread, shoemaker's thread, fishing lines and nets, carpets, fire-hoses, bed-linen, and tablecloths. The finer grades are used for handkerchiefs, lace, and embroidery.

Jute

In contrast with fine fibres like flax and cotton, jute (which takes its name from a Sanskrit word meaning "matted hair") serves the workaday needs of commerce and industry. It comes from a tropical plant (*Corchorus capsularis*) that grows on the flat flood plain of the river Ganges, around Calcutta. Here peasant cultivators sow it densely, weed it carefully, and harvest it after it has grown eight to ten feet tall. It forms slender, upright stems bearing simple leaves and small, yellowish flowers. Jute is harvested before the seeds form, and the stems are retted in bundles sunk in the watercourses. After soaking for three weeks they are raised and beaten hard on the surface of the water. This frees the soft tissues, which are washed away, leaving the strong bast fibres behind. About 1500 pounds of dry fibre are obtained from one acre of the crop.

Nearly all the jute harvested goes to two cities, Calcutta in India and Dundee in Scotland, where great industries have developed to process it with complex machinery. It is a strong yet coarse fibre, which will not yield an attractive white cloth. So it is used for cheap sacks, for wrapping cloths such as burlap and hessian, and as a backing fabric for linoleum; smaller quantities are used in the manufacture of paper.

The jute-plant may grow to a height of at least 10 feet and yield individual fibres 8 or more feet long. Like flax, jute is retted after harvesting.

Raw jute (above, left) is taken to the mill (above), where it is woven into rolls of coarse cloth.

Hemp

Hemp is a tall annual plant (*Cannabis sativa*) with compound leaves. Native to Asia, it has been grown from very early times in China (and since the Middle Ages in Europe) as a source of fibres for making ropes. Grown from seed, it is cultivated in America, Russia, Poland, Italy, and China, but many countries prohibit its cultivation because (as we have seen) it is the source of the drug hashish, or marijuana. Its round, grey seeds are used for feeding cage birds, and as a source of oil. Hemp is a stem fibre and is harvested and prepared like flax. It is not flexible enough to weave into cloth, but can be spun into very strong and durable ropes and cordage.

The hemp-plant is harvested immediately it flowers, when it is about four months old and five to eight feet high. Hemp was cultivated by the Chinese at least 4000 years ago.

Abaca or Manila Hemp

This important fibre, which has replaced hemp to a large degree because it has greater strength for a given weight, comes from a completely different plant (*Musa textilis*). The abaca plant is closely related to the banana, and produces, like that plant, huge false stems that are really leaf stalks springing up from an underground true stem. These tall upright stems are cut down just before the plant is ready to flower. All their soft tissues are scraped away by working the stems between a sharp knife-blade and a wooden block, or else by machinery. The remaining soft, white fibres, which are really the veins or vascular bundles of the leaf stalks, are then dried.

Abaca, which first attracted attention as a commercial fibre as recently as 1820, is native to the Philippines, but is now also cultivated in Indonesia and Central America. It is raised from offshoots of the rhizomes, which give rise to fresh crops at intervals of 12 to 18 months.

Sisal

Sisal, a typical leaf fibre, consists of the veins or vascular bundles that form part of the leaves of the sisal plant (*Agave sisalana*), a native of Mexico. The plant is adapted to life in a dry climate on the edge of deserts, and develops a huge rosette of fleshy leaves. After some years it grows a large flower stalk, up to twenty feet high, that bears little buds called *bulbils*, as well as seeds. The bulbils, or alternatively cuttings from the plant's rhizomes, are used to start a new crop. The plant grows slowly, but after four years the huge leaves on the outside of each tuft-shaped plant begin to bend outward, and are ripe for cutting. They are harvested with sharp knives, and from this time onward a fresh crop can be cut each year, for they are

The sisal-plant bears thick leaves on a stumpy stem. The fibres are taken from mature leaves that may be six feet long.

replaced by fresh leaves that initially grow upright at the heart of the plant. After about eight years of cutting, the sisal plant develops a tall-growing inflorescence and then dies.

Only a small proportion of the leaf consists of fibre; the rest—the soft pulp—has to be stripped away. Nowadays this is done by machines that crush the leaves (but not the tough fibres) and wash the pulp away. Sisal fibres are about three feet long, rather coarse, but very strong. Their main use is as twine for packaging, especially as binder twine used in grain harvesters. The crop is grown in East Africa and Indonesia, as well as in its native Central America and the West Indies.

Sisal fibres emerge (above, left) from a decorticator on a plantation in Kenya. The machine crushes the leaves and scrapes away the pulpy tissues. The cleaned fibres are washed, centrifuged, and (above) dried in the sun.

Individual strands of sisal fibre are fed (below, left) into a rope-making machine. Below: a reel of sisal cord.

Reeds, Grasses, and Palm Leaves

In many countries of the world extensive use is made of the long straight stems of certain plants as building or roofing materials. Most of the plants so used are Monocotyledons, which usually have long slender stems and leaves bearing few or no branches. Many of these plants are grasses of the family Gramineae, which includes the grain crops; others belong to the reed family (Juncaceae) or to the sedge family (Cyperaceae). A common feature of all these plants is a hollow stem, which may be round, triangular, or flat in cross-section. The hollow centre enables the cut stem to shed water easily and to dry quickly after wetting. The grouping of the hard strong fibres around the outside of the hollow makes the reed flexible and easy to work.

Reeds and grasses for thatch are gathered after the stems have grown tall through the summer and are beginning to wither. As a rule they are sun-dried so that they wither quickly and natural decay is checked. Thereafter, they will resist decay for many months or even many years provided they are used in situations where they do not collect water and can dry out quickly after getting wet.

Thatch has been used as roofing material in England and other European countries for many hundreds of years. If it is made of wheat straw it will last for 10 years, but if it is made from the great reeds that grow in fresh-water marshes it may endure for 30 years or more. The hollow stems are cut in the autumn, their leaves stripped away, and they are made up into bundles. Each thatched roof has two layers of thatch —an inner one that is fastened with hazel pegs to the wooden rafters of the house, and an outer one that is pegged on to the inner one. The combined layers are often at least a foot thick. The materials used are cheap, but thatching is slow work and requires the expensive skills of a craftsman, and therefore few new buildings are thatch-roofed today.

In many tropical and subtropical countries where there is no need for houses to be very warm, grasses and reeds are often used for walls. Grass huts do not last for long, but the material is so cheap and so easily handled with simple tools that this is not a serious drawback for people who are ready to do all the work themselves. The Marsh Arabs of Iraq, who live on the flood plain between the lower reaches of the Tigris and the Euphrates, make their homes and storage buildings out of huge bundles of reeds. This is indeed the only suitable building material hereabouts. There are no trees, and therefore no timber, on the flood plain,

Two widely scattered examples of the thatcher's craft: a chief's house on the Trobriand Islands, New Guinea, and a row of cottages at Ringwood, Hampshire.

and the land is too soft and too often flooded to bear buildings of brick or stone, even if the Arabs could afford them. But reed buildings are easily and cheaply built and replaced.

In tropical Asia the characteristic roofing and thatching material consists of broad, fan-shaped palm leaves, which are woven loosely together so as to overlap. In Malaysia, the resulting screens, called *attaps*, keep out wind, rain, and strong sunlight, and ensure privacy; but air can filter through them and so a hut remains cool, dry, and well-ventilated. When laid more thickly as a roof-cover, the leaves keep out the fiercest tropical rains.

Construction of an Arab sheik's *mudhif,* or guest house, in the marshlands of southern Iraq. Two rows of massive reed bundles are embedded in the ground (upper left), then are bent inward (upper right) to form a series of arches. The skeleton of the building is completed by cross-ties to make the arches rigid, and then the whole exterior is covered by securely bound reed matting. The lower picture shows the pleasantly cool, dark interior of the *mudhif.*

Basketry and Brush Fibres

The two main sources of fibres for basketry are also the grasses and the palm trees. A great range of different materials is used in various countries. Most have only a local application, and only a few enter world trade. Along the coasts of north-western Europe, for example, there grows a tough, hard grass called *bent grass* or *marram*, which thrives on sand; its wide-spreading roots help to maintain protective coastal dunes by binding the sand particles. The stems of marram can be woven into baskets and mats, and even shaped into sturdy stools or chairs. Raffia, which is widely used for ornamental baskets, consists of leaf strips of the raffia palm (*Raphia ruffia*), which grows in Madagascar. Panama hats are woven from strips of leaf cut from a climbing, palm-like tree that grows in South America. Cane, or rattan, used in wickerwork consists of the long slender stems of bamboo-like jungle climbers (mainly of the genus *Calamus*) that grow in Malaysia and elsewhere.

The leading basketry material in temperate lands is willow (genus *Salix*). Selected strains are grown from cuttings on fertile land and are cut back to the ground as soon as they are established. This causes the stump to send up many slender, flexible shoots, which are harvested after only one year's growth, when they are about six feet long. New crops can be cut for many years. The thinner strands are made by splitting the rods lengthwise, using a special hand tool.

The tough fibres used in brooms are known by their Amerindian name, *piassava*. They are cut from the base of the leaf stalks of a wild-growing palm tree (*Attalea funifera*) native to eastern Brazil, or from a similar species (*Raphia vinifera*) that grows in tropical Africa. (The Brazilian species yields the coquilla-nut, the hard shell of which is used in wood-carving and turnery.)

The raffia palm, a native of Madagascar.

Below, from left: rattan basket (Borneo), raffia beer-strainer (Uganda), grass milk-pot (Somalia), coiled-grass basket (Kenya), eskimo grass hat, depicting a whale hunt (north-west coast of America).

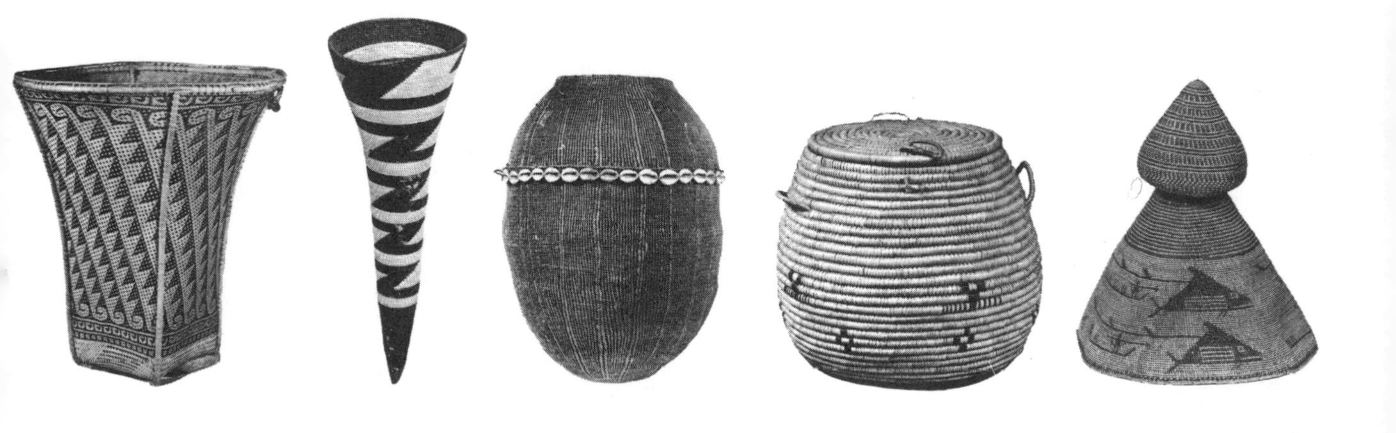

Cork

As we saw in Chapter 1, under the bark of all trees there are two growth-layers called *cambia*, one beneath the other. The outer layer, the cork cambium, produces bark; and it carries out this job steadily throughout the life of the tree, because the trunk is always getting stouter. The inner layer, the wood cambium, produces both new wood on its inner side to conduct sap up the tree, and other conductive tissue on its outer side, called bast, which carries sap down the tree. If we strip the bark from a tree, it usually breaks away from the inner, wood cambium, and the bast comes away with it. As a result, the tree will die.

The cork-oak (*Quercus suber*) is different from all other common trees, including related oaks. If we strip away its bark, it breaks away at the outer, cork cambium. The bast below is undamaged, and the cork cambium gradually grows a fresh coating of bark to protect the tree. This unique feature makes it possible to harvest repeated crops of bark from the cork-oak about once every 12 years.

Cork is a waterproof substance that protects the trunk from insects, fungal diseases, and chance wounds, and shields the bast from the hot rays of the sun and from short spells of sharp frost (though it cannot eliminate the effects of long-sustained winter cold). The cork-oak, which is native to the south of Spain and Portugal, has an exceptionally thick bark because it grows in a region of long, hot, dry summers: to keep alive, it has to hold what moisture there is firmly within its trunk. For the same reason, its leaves

Cutting the outer bark from a cork-oak tree (above) and transporting it (right) to the market in the Algarve, southern Portugal. The several hundred thousand tons of cork produced every year come mainly from wild-growing trees in Spain, Portugal, and Algeria.

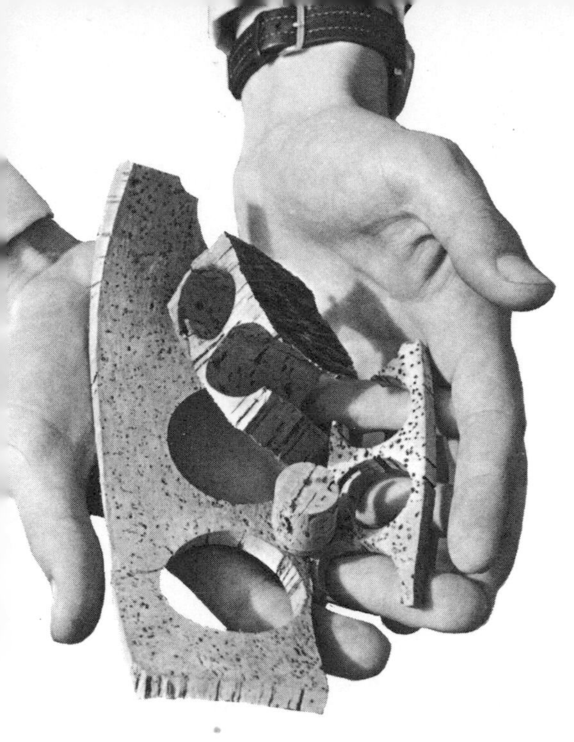

Unprocessed cork has been used to make these simple bottle stoppers.

Below: the cork used in lining metal stoppers (right) is first crushed to powder, then mixed with resin and compressed under high pressure into thin sheets for stamping; gaskets are also made in this way. Flooring cork is powdered and then mixed with liquid rubber (left) to make it flexible. The mixture is rolled into sheets and cut to the required size.

are hard and glossy; they are evergreen, because they get enough water for active life only in the cooler winter months.

Cork had little value until vintners began to use glass bottles to hold wine. Then it was discovered that if cork was handled in a special way it could be made into excellent stoppers that would prevent the volatile alcohol escaping and would seal off the wine from the atmosphere. Every piece of cork is threaded through with parallel breathing pores to enable the tree trunk to get air. But if a bottle stopper is cut lengthwise from a thick sheet of cork, the pores run across the neck of the bottle, and are sealed off by the glass around it. The water vapour and the alcoholic spirits, which would otherwise evaporate from the wine, are stopped by the thick, continuous layer of impervious cork. Today, however, the most important uses for cork are in the manufacture of sound-insulating materials and in flooring such as linoleum.

Nearly all the world's cork comes from long-established groves of cork-oak on estates and peasant holdings in southern Portugal and Spain and from Algeria. The first crop from young trees, which are raised from seed, is called virgin cork. It is rough in texture and unsuitable for stoppers, but is used, along with the waste from finer grades, for heat insulation in refrigerators or as a floating material for net floats or seamen's life-belts. This and the succeeding crops are harvested by men who peel off sheets of bark with sharp, curved knives. The bark is simply dried in the sun, then sorted and graded for export.

14 Forests, Timber, and Paper

Above: thousands of logs, gathered into a solid mass by an enclosing boom, are towed to a sawmill in Newfoundland.

Wood was the most important raw material known to early man. His progress toward civilization depended, at least in part, on his ability to fashion it into tools, weapons, household goods, buildings, and fences; it also served him well as fuel—to keep him warm, cook his food, fire his potteries, and smelt his metals. Although man now uses a variety of other materials for jobs once done mainly by wood—metals for tools, for instance, coal, oil, and gas for fuels, and concrete for building—he still needs ever-growing quantities of timber, both for traditional uses and for new applications developed by the modern technologist.

Early man took the forests very much for granted: nobody had planted them, and they were so vast that there seemed no limit to their rate of growth and renewal. For early agriculturalists the forests were, as we have seen, a barrier to progress: they had to be destroyed to make the land available for grain and other food crops. It was only when man began to need timber in large quantities that he came to realize that forests demand as much care as any other form of plant life. Trees are a crop and must be treated as such.

Forests not only provide raw materials; they also help to prevent soil erosion, especially on hill and mountain slopes, and give shelter to crops and stock by acting as a wind-break. By moderating the flow of rain-water from the land, they help to prevent the recurring cycle of flood and drought common in many treeless areas.

We saw in Chapter 2 (pages 38–9) that, if undisturbed by man or by climatic or other changes, vegetation tends to progress toward a climax, the nature of which is determined, from region to region, by climate and soil. Between the tundras of the Northern and Southern Hemispheres, climax vegetation is typified by forest species of various kinds (though the forest is interrupted at many points by grassland, desert, and areas under human cultivation). At and near the equator, where temperatures are always high and rain falls heavily throughout the year, there are tropical forests

Below: illustration from a scientific work on forestry and timber production published in France, 1764. In the conical hut in the background, wood is being burnt for charcoal; at left are surveyors; on the platform at right three men cut a massive plank with a special saw.

containing a remarkable variety of species of trees, shrubs, and climbing plants. Almost all are broad-leaved evergreens, because there is no marked seasonal climatic change that might cause them to shed all their leaves at one time.

Beyond the tropical forests are regions that experience a dry or a cold spell (and possibly both) at least once a year. This limits the number and range of species of trees that can thrive in such regions, and the forests are usually more open; the savannah forests, for instance, include many evergreen palm trees spaced far apart.

In the temperate zone, broad-leaved forests are found again. But owing to the cold winters that are a feature of the zone, the trees are mostly *deciduous* (from the Latin *decidere*—to fall off): they lose their leaves every autumn and do not grow a fresh crop until the following spring. Still farther north are the forests of evergreen coniferous trees, whose needle-shaped leaves are able to survive the coldest winters.

Conserving Natural Forests

Nearly all trees start life from seeds that develop from flowers borne high in the tree-tops. A few, however, spring up by vegetative reproduction from the roots of their parents, and some grow from cuttings prepared by man. In a natural virgin forest there is a slow but steady succession of young trees growing near, and competing with, older ones. When an old tree dies and falls, the gap that is left is soon filled in by seedlings and saplings. It may take a tree 200 years or more to reach maturity, but all the time it is surrounded by other trees of various ages and sizes, down to the smallest ones sprouting from newly fallen seeds. The forest, then, is a dynamic system, constantly renewing itself by fresh growth.

In the early years of the 18th century, foresters in Germany realized that, if they carefully chose the number and age of trees to be felled each year, they would do no lasting harm to the forest. Natural regrowth would make good the felling loss at no cost, and they could take a small but steady annual harvest from the woods for as long as they wished. This is the principle of *conservation* of forest resources, combined with *sustained yield*, which guarantees a steady income to the forest owner and a steady supply of timber to the people who use it. The application of these ideas has now spread to all parts of Europe, to India, to North America, and to many other countries that have extensive natural forests.

Where the natural forest cover has been destroyed, people have to start at the very beginning and raise trees from seed or cuttings. As early as 1460, a Scottish monk, William Blair of Coupar Angus Abbey in Perthshire, was tending nursery beds of forest trees to plant on the abbey's estates on the neighbouring hills. This is one of the earliest examples in Europe of man deliberately cultivating trees for timber.

In some countries tree seed can be sown directly on the land where the forest is to grow. But as a rule the best method is to sow it in a nursery. Often the seedlings that spring up are too small to be taken straight to their final home; if they were allowed to grow unchecked, they would not produce large enough roots to enable them to survive in the plantation. So they are first transplanted to another nursery bed. This treatment checks the growth of their shoots, promotes the growth of their roots, and leads to a well-balanced plant that will thrive well when it is planted in the forest later on.

Nurserymen raise many thousands of small trees, and these are usually planted out in the cold or the

Above: a seed-bed of six-month-old spruce trees grown from seed. A year hence they will be transplanted to another, less-crowded nursery bed.

Below: a transplanting machine at work. The operators place the seedlings, roots upward, in clips on turning wheels that plant them at the correct intervals as the machine traverses the nursery plot.

Above: young spruces two years after transplanting. They have now developed bushy root systems and are ready to be planted out in the forest.

Below: brashing (removing lower branches of) spruce-trees leads to cleaner timber with smaller knots, lessens the fire risk, and provides the forest soil with a mulch of nutrient-rich brushwood.

rainy season. The land used for new forests is usually unsuitable for other forms of agriculture; it includes steep mountain-sides, peat bogs, and poor sandy soils that may need terracing, draining, or ploughing before planting. If the soil is especially sandy or peaty, and so lacking in sufficient nutrients, it is usually fertilized, especially with phosphates.

The young trees are spaced about five feet apart, and one worker can plant 500 in the course of a day. Fences are nearly always needed, for farm livestock, rabbits, or deer are always ready to bite back the young trees and the grass growing around them.

If all goes well, the trees grow steadily taller and spread out until their branches meet and they form a thicket. In the colder regions this may take ten years, in the tropics only three or four. Henceforward the trees compete with each other for the available ground and sunlight. After a few more years, the forester takes a hand by thinning out his crop. He leaves the better trees to grow, and cuts down some of the others for sale. He repeats these thinnings every few years, because the survivors always expand into the available space. At last, when the trees have grown as big as anyone wants, the forester fells his final crop. Sometimes the cleared ground is readily restocked by seedlings springing up from seed shed by the fallen trees; but as a rule it is necessary to replant the land with young trees raised in the forest nursery.

The age of the final-crop trees varies a great deal according to the region and the species of tree. In subtropical lands such as South Africa, pine crops may mature at 30 years, but European foresters growing oak may keep the best trees until they are 200 years old. In the meantime, of course, they will have harvested many lesser ones, thinned out from the main crop. Foresters who look after planted crops must still pay regard to the principles of conservation and sustained yield. They divide their woods into sections, each containing trees of different ages, so that planting, thinning, and final felling go on continually over the years.

All foresters are constantly on their guard against fire. In dry weather, both the trees and the vegetation around them become very inflammable. So each forest is broken up by fire-breaks, and men keep watch from lookout towers for any signs of smoke. Fire-fighting lorries, equipped with motor-pumps and water supplies, are kept ready to rush men to any blaze. In North America, aeroplanes and helicopters are used also, and fire-fighters (called "smoke-jumpers") are sometimes dropped by parachute.

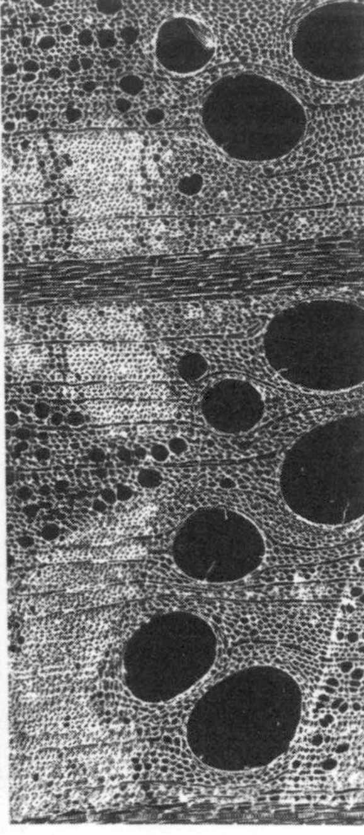

Transverse sections (above) and radial sections (below) through the trunks of two trees: on the left, softwood (Scots pine, × 30); on the right, hardwood (English oak, × 50). Note annual rings in both transverse sections.

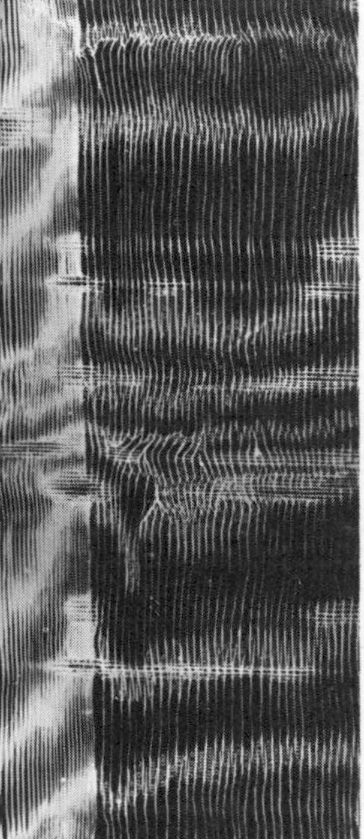

Timber and its Uses

The wood harvested from the stems of forest trees is one of the most complex and adaptable raw materials on earth. In the living tree, it serves two purposes: it carries sap upward from the roots, and it supports the weight of the tree's branches and foliage, holding them firm against the enormous pressure of the wind. In order to do both these things, wood is built up of millions of long narrow cells with fibrous walls. The cell walls contain two substances: *cellulose*, which is very flexible, and *lignin*, which adds strength and stiffness. In the living tree, the hollows in the cells are full of water. After it has been felled, the water evaporates in the process of seasoning. As a result the wood loses about half its weight; it is, in fact, one of the strongest substances known, in relation to its weight. Moreover, it is easily cut or shaped with hand tools or powered machinery.

Wood is a good fuel, and millions of people who live in or near forests still rely on it for winter warmth and for cooking their meals. If it is partially burnt in a kiln that restricts the access of air, it is converted into charcoal. This is a form of almost pure carbon that burns at high temperature when given ample air. Charcoal is still used by many people for cooking, and, until the invention of coke, charcoal was the only effective fuel for smelting iron and other metals.

Another very ancient but still important use of wood is as the handles of tools and weapons. Only a few species of timber are sufficiently strong and elastic to stand the force of a man's muscles striking repeated blows or levering weights. Two of the best are ash (genus *Fraxinus*), which is widely used in Europe, and hickory (genus *Carya*), which is more popular in North America. Vast tonnages of wood are used in construction, mainly in industrial buildings, houses, wharves, bridges, and fencing, and in furniture and interior fittings. The chief disadvantage of wood in these applications is that it is attacked by certain insects and fungi, which use it as food. The common furniture beetle (usually called "wood worm") attacks indoor and outdoor timber, especially if the wood is damp. The death-watch beetle also thrives in damp conditions, and attacks timber whose surface has been softened by fungal infection. The death-watch beetle is an especially serious pest of hardwood timber, such as the beams and joists in old buildings. In the tropics —and also in some temperate regions—the most serious timber pests are the termites, some species of which thrive on damp wood while others favour sound, seasoned wood. With suitable protection from

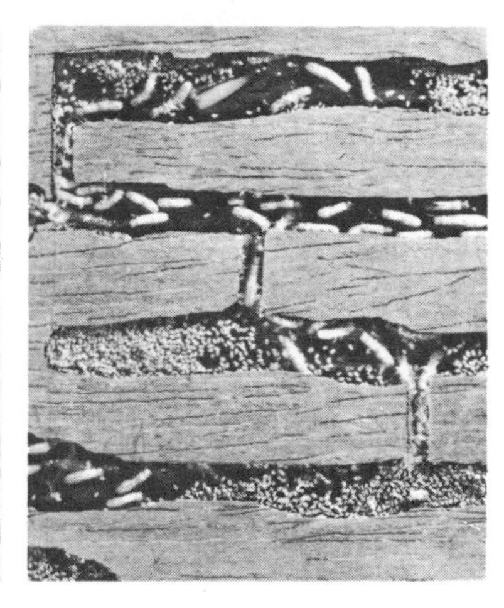

Common enemies of wood: fruiting bodies of dry rot (left) caused by leakage from water pipes; common furniture beetles (centre) at exit holes they have bored in plywood; West African termites (right) in galleries bored in plywood.

damp, timber will endure indefinitely. Oak beams hundreds of years old still support cottages and churches in many parts of Europe, and wooden coffins have remained sound for thousands of years in the tombs of Egyptian pharaohs. Today timber can be protected against fungi and insects with chemicals, which are injected deep into the wood under pressure.

Plywood, Man-made Boards, Paper, and Rayon

As we have seen, many of the traditional uses of timber—especially in tool- and furniture-making and in building construction—depend upon the strength, elasticity, and workability that are inherent in felled seasoned wood. The interesting thing about many of the newer uses of timber is that the wood is deliberately broken down into thin strips or into pulp, and is then built up again in layers or is reconstituted in a totally different form. Plywood, for instance, is made by peeling thin layers, or *plies*, off a log and then gluing them together in such a way that the new material has great flexible strength in all directions.

Below, left: production of veneer. After its bark has been removed, the log is rotated against a lathe, which peels off a continuous sheet of wood. Below, right: in plywood manufacture, several sheets of veneer are placed on top of one another, glued, and then fed into a hot-press that bonds them securely.

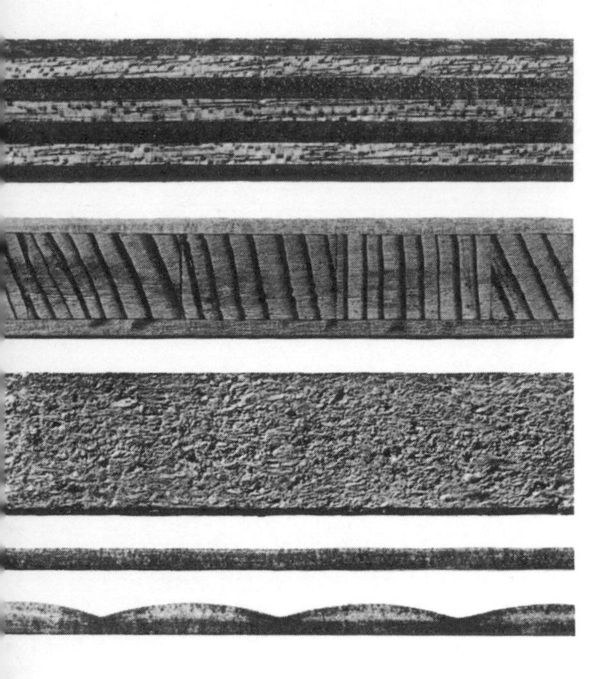

Above: four types of reconstituted wood. From the top: plywood, laminated board, chipboard, and two examples of hardboard, one with a wavy surface.

As a rule, wood from different species of trees is used for adjacent plies. In *laminated* boards and beams the same principle is used to make stouter and longer units. *Chipboard* is made by cutting up logs into chips, and then gluing the chips together, under heat and pressure, with plastic resins. *Hardboard* and *insulation board* are made by breaking down wood to its individual fibres and then reuniting them in a new pattern. These last three materials are cheap because they can be made from wood waste and from small logs that otherwise would have little commercial value.

Paper was invented by the Chinese about 2500 years ago, and by the Christian era they were using paper money as official currency. But this paper, like almost all others until the 19th century A.D., was made from fibres other than wood. The Chinese commonly used cotton or linen rags as raw material, as did many other countries, and rags are still the basis of high-quality writing papers. Other common materials were the fibres of grasses, including esparto grass and bamboo.

The use of wood pulp—mainly spruces, balsam firs, hemlock, poplar, aspen, and Scots and several other species of pine—was due essentially to increased demands for paper, which could not be met from existing raw materials. The demand reflected generally the rise in population in Europe and North America and specifically the increasing literacy of the peoples of both continents. The use of wood pulp, and the development in 1798 of a machine for making continuous lengths of paper ultimately made possible the mass-circulation newspaper.

Below: in chipboard production these slivers of timber will be mixed with hard-setting synthetic resins and shaped into boards under heat and pressure.

General view inside a hardboard manufacturing mill. The man seated in the background manipulates a panel that controls the hot-press and conveyors of the production line at left. In the production line at right, light-coloured insulation board is about to enter the slotted drying tunnel.

Paper-making (above, left) in China in 1634, and (above) in 18th-century France. In both, a vatman plunges a mould into a vat of fibres and water. A sheet of paper forms when he withdraws the mould, sieves out the water, and shakes the mould so that the fibres interlock.

Paper-making now uses 40 per cent of the world's timber harvest, North America, Europe, and the Soviet Union providing most of the timber required for this purpose. The cheaper, less-durable types of paper, such as newsprint, are produced by a simple process. The logs are ground, by massive grindstones rotating under water, in such a way that their fibres are separated but not damaged. The resulting mixture of fibres and water then passes on to a moving belt of wire mesh. The water drains away and, as it does so, the fibres become interwoven into a thin fabric, which is dried and smoothed over hot rollers and emerges as a continuous roll of paper. The whole process, from wood through *groundwood* or *mechanical pulp* to paper, takes less than an hour.

For higher grades of paper, the timber is first broken down into chips and then *digested*, or cooked, with powerful chemicals under pressure. These processes dissolve the lignin, and the resulting *chemical pulp* is almost pure cellulose. The rest of the basic paper-making cycle is similar to that for groundwood, except that the pulp passes through a series of fine screens to remove the coarser fibres; in addition, the paper is bleached and treated in a variety of other ways to give it specific properties such as strength, smoothness, fineness of texture, and water-repellency.

If chemical processes are taken a stage further, the cellulose in the wood pulp can be dissolved and then reconstituted as the very fine silk-like thread called rayon or viscose, which is widely used in the manufacture of stockings, shirts, dresses, carpets, and tyres.

Below: in rayon production a solution of cellulose is drawn through the holes in the spinneret and instantly solidified into threads by hardening chemicals.

1

2

The photographs on these two pages show eight of the principal stages in the manufacture of paper from wood-pulp.

(1) Pulp-wood logs are piled high in this New Hampshire yard, where they await processing at the mill in the background.

(2) At the mill the logs are debarked in revolving drums in which they are tumbled together under powerful jets of water. In some mills the bark is dried and used as fuel in the boiler house.

(3) Logs are reduced to pulp in this shop by pressing them with hydraulic rams against water-cooled grindstones revolving at high speed.

(4) This 40-foot-high chemical digester dissolves the natural binding substances out of the wood, leaving the pulp fibres.

3

4

5

6

(5) Sheets of dried pulp are pressed into bales for processing elsewhere.

(6) At the paper mill the dried pulp is placed in this hydropulper and swirled around with water at high speed until it resembles a thick porridge. Groundwood and chemical pulps are mixed in varying proportions according to the type of paper required; chemicals are added to give the paper a good printing surface.

(7) In the paper-making machine the fluid pulp flows onto a moving sieve, through which most of the water drains, then is carried on a felt pad through rollers, which squeeze out more water, and finally is dried out against a series of steam-heated cylinders.

(8) The paper is calendered by passing it through a series of heavy rollers that impart a gloss to its surface.

7

8

The Conifers of the North

The world's richest store of timber is the vast coniferous forests that girdle the Northern Hemisphere, through Scandinavia, the Soviet Union (including Siberia), Canada, and the northern United States. There are many outliers on high mountain ranges farther south—the Alps, the Caucasus, the Himalayas, and the Sierra Nevada of California. Most of the trees are evergreen, cone-bearing species that belong to the natural order Coniferae. They are also called *softwoods*, because their timber, though strong enough for most purposes, is relatively soft and easy to work: it is in much greater demand for constructional use and packaging than the *hardwood* from broad-leaved trees. Moreover, it is easier to make into paper because its fibres are longer.

The typical Scandinavian conifer forest is dominated by various species of pine and spruce, together with lesser amounts of birch. Harvesting takes place in the late autumn: the trees are felled by axe and power saw, and their tops and side branches are lopped off to make the massive stems easier to handle. As in Canada and other countries, Scandinavian foresters use the complex system of interconnected lakes and rivers to transport the timber from forest to mill. By the time all the selected trees have been felled, the snow is usually thick on the ground. For the short journey from forest to lakeside, one end of each log is hoisted on to a sledge drawn by tractor or horse.

The logs are piled on to the lake ice, where they

In northern latitudes most of the timber is felled during the winter and hauled to the nearest river or lake to await the spring thaw. When the ice melts rivers become raging torrents and carry the logs swiftly downstream toward the mill.

Lumber mill on Vancouver Island, Canada. Most of the timber here is destined for construction work. In the foreground stacks of planks are being seasoned.

Forest fires pose a greater threat to timber supplies than any other hazard—hence these fire beaters placed at intervals throughout an English forest.

remain until the spring. With the spring thaw the logs begin to move on their slow journey downstream toward the mills. On arrival they are sorted, while still afloat, into various grades according to species and size. Some are fed into a sawmill, where they are cut into planks and left to season in the open air. Others go to the paper mill; still others to the cellulose mill, where they are converted into the high-grade pulp used in rayon manufacture.

The most magnificent forests of the northern conifer zone stand on the western coast of North America, where the Rockies slope in a network of river valleys toward the Pacific Ocean. Some are in southern Alaska, some in British Columbia, and others in the American states of Oregon, Washington, and California. In these forests, many individual trees soar to a height of 300 feet and more. The California redwood (*Sequoia sempervirens*), a tree with thick, soft, red bark, holds the record for height: the tallest so far discovered, near Redwood Creek, northern California, is almost 368 feet high. The allied American Big Tree, or wellingtonia (*Sequoia gigantea*), which has scale-like leaves, is the world's stoutest tree; one veteran, called General Sherman, has a maximum girth of 75 feet. Counts of the annual rings of similar monarchs after felling have proved several to be about 3000 years old; it takes at least 500 years to produce even an average-sized redwood. But these enormous trees form comparatively small groves, and many are preserved from felling in national parks. In terms of tonnage felled,

Below: diagram showing annual production of paper and paper board in the United States. Note the importance of waste-paper pulp as a raw material. (A cord equals 128 cubic feet, stacked measure.)

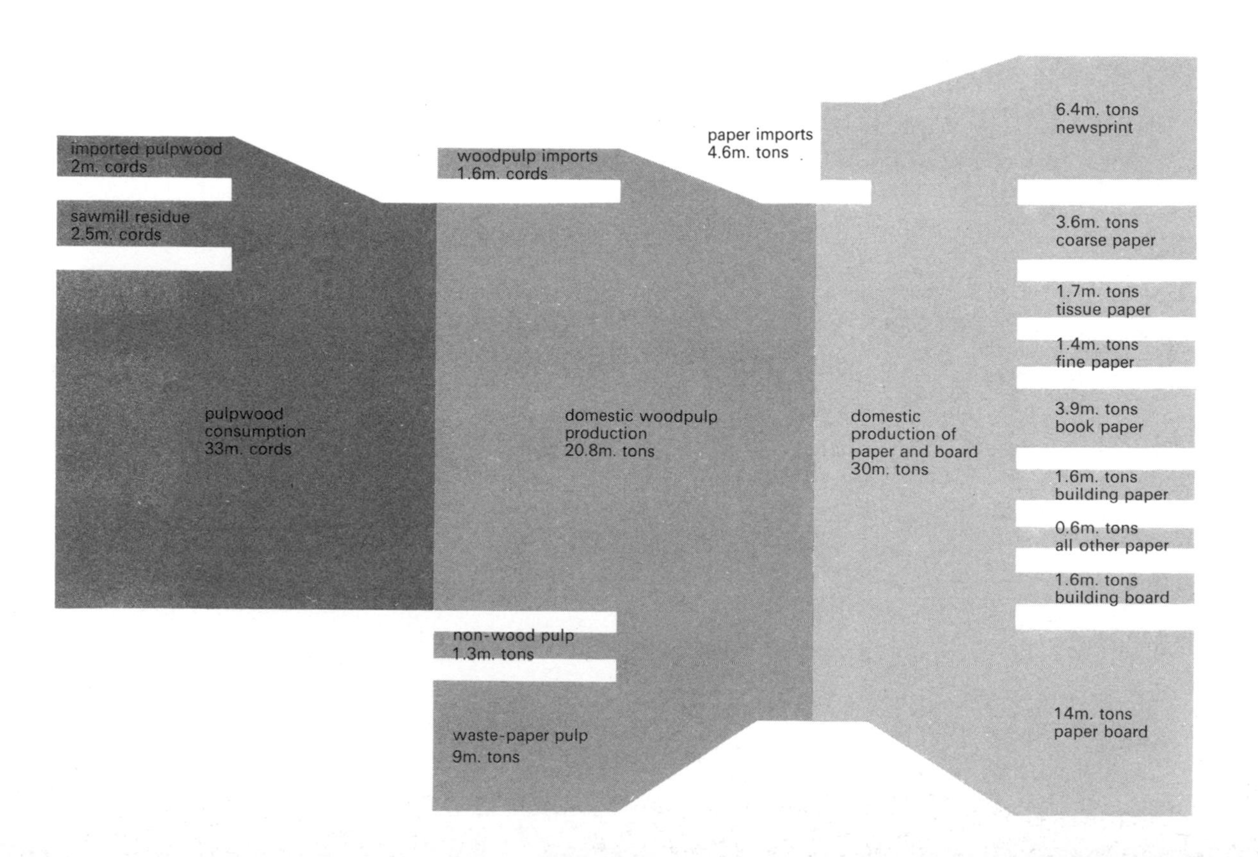

the main timber species are the Sitka spruce (*Picea sitchensis*) and the Douglas fir (*Pseudotsuga menziesii*), with lesser amounts of lodgepole pine, western red cedar, Lawson cypress, and noble fir. The North American conifer forests are so dense, and carry so much timber, that the native Indians made few clearings, and lived by fishing and hunting rather than by agriculture. Only in the late 19th century, when the Europeans brought in machinery to handle the huge logs, did it become worthwhile to fell them. Most of the forests belong to the governments of the western states, which lease large areas to timber companies under strict conditions; though the companies may carry out large-scale fellings, they must leave seed trees to provide adequate seed for the next crop.

The forest giants pose a tricky handling problem. The first job is to bulldoze a rough but firm track between the felling area and the mill. At the side of the track, in the centre of the felling area, a huge steel tower is erected and fitted with powerful winches and cables. The great trees are felled with petrol-driven chain saws; then their branches are cut away, and their trunks cross-cut into sections. Each section, weighing several tons, is winched in to the steel tower. Here a device like a giant crane, called a *yarder*, picks up the great logs and loads them on to trucks, which deliver them to the mills.

To the south of the coniferous zone are the forests of broad-leaved, deciduous hardwood trees. Ever since the beginnings of agriculture, these forests have suffered more than any others at the hand of man. The reason is that these trees are deep-rooters: they produce a considerable depth of soil and, moreover, enrich it every year with a heavy fall of humus-forming leaf. The resulting brown forest soils are highly fertile and a great attraction to the farmer.

The broad-leaved forests of Europe, Asia, and North America are much alike: the dominant trees in each continent are all of the same genera, though there is considerable variation in species. Oaks (mainly genus *Quercus*) are abundant and yield very strong timbers, with durable heartwood, that are used for house-building, ship-building, furniture, and fencing. Beech (genus *Fagus*) and various species of sycamore and maple (both genus *Acer*) give good furniture timbers. Ash, as we have mentioned already, is used for tool handles, and also for oars, hockey sticks, and the framework of motor vans. Although these trees seldom grow very fast, selected hardwoods often fetch high prices because they are difficult to replace by other species or alternative materials.

California redwood

Sitka spruce

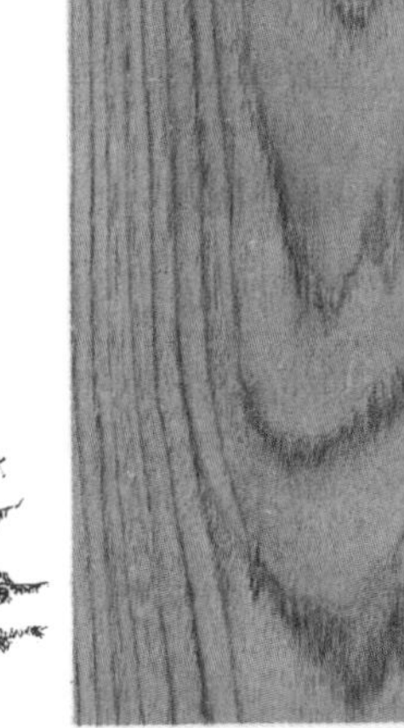

Douglas fir

Pedunculate oak

Drawings and photographs at right show eight examples of commercially valuable trees and veneers of their timber.

Man-made Forests in the Temperate Zone

Many countries that have lost their natural forests through overcutting are today engaged in ambitious re-afforestation schemes. A curious feature of many such schemes is the cultivation of *exotic* species—that is, species indigenous to other regions that, for various reasons, thrive better than local species. A good example of this is the new man-made forests that are developing in hot, dry regions of such countries as Spain, Israel, East and South Africa, Australia, Argentina, and Chile. Forestry commissions in all these countries have imported several species of pine from California and the south-eastern United States. One such species is the Monterey pine (*P. radiata*), a rare tree found on the California coast, where its appearance is wind-swept and somewhat stunted compared with other pines. In its new habitats, however, the Monterey pine grows fast, straight, and tall, and yields strong sawmill timber by its 20th year. This is especially remarkable because many trees are difficult to raise successfully in these arid lands. Hillside plantations are usually terraced to hold the limited rainfall, and pines are often raised in pots before being transplanted.

Eucalyptus trees from Australia have also been planted on a large scale in Spain, as well as in Italy, Greece, and North Africa. They grow very fast and yield timber suitable for fencing, fuel, and paper pulp. In the British Isles the most profitable trees for planting are the Sitka spruce and the Douglas fir from the Pacific slope of North America. On the western seaboards of Scotland, Wales, and Ireland these trees find a climate very like that of their homeland, and grow faster than any similar species native to the temperate zone of Europe.

In China there is a vast programme of afforestation which aims to clothe the steeper hill-sides with trees, partly for timber production and partly to stop erosion of the soft *loess* soils. The loss of these soils from the hills, which has continued unchecked for centuries, fills the rivers with silt, and in the rainy season causes disastrous floods that often destroy the paddy fields. Once the forests are established, the trees will bind the soil, reducing the runoff and checking erosion.

On the steppes of Russia, "shelterbelts" of trees, many miles in length, are being planted in a criss-cross fashion to moderate the fierce winds that sweep across the hitherto treeless plains. These belts will check both the fiercest winter blizzards and the dry summer gales, and make a better environment for crops, livestock, and man himself.

Tropical Jungles

The evergreen broad-leaved forests, or tropical jungles, are of less commercial value than the forests of the colder and drier lands. For, while they contain many valuable trees, these are surrounded by so many others of little or no value that it is difficult to harvest the good ones profitably. The local people get most of their timber needs cheaply from the smaller and more easily worked trees, and leave the felling of the forest giants to European and American companies that ship timber overseas. In the past, only man-power and animal power—oxen, buffaloes, and elephants—were available to move the mighty logs; even today, elephants are still widely used in India and elsewhere, though they are now giving way to mechanical power.

Teak (*Tectona grandis*), which grows in the monsoon rain-forests of southern India, Burma, Thailand, and Indonesia, was early recognized as one of the world's strongest and most durable timbers, and of particular value in ship-building. But only a few good, mature teak trees grow in each square mile of forest, and each is surrounded by hundreds of other relatively valueless species. When a marketable teak tree is found, it is

Cutting (left) and hauling away (above) a trunk of sapele (*Entandrophragma cylindricum*), sometimes called African mahogany, a valuable hardwood that grows in the tropical forests of Ghana and other parts of West Africa. The straight trunk, which may be 30 feet in circumference, may grow to a height of 100 feet and more.

Above: West African timber is impregnated under pressure in a sealed cylinder with insecticides and fungicides before being used in construction work.

Below: sapele logs in Takoradi harbour, Ghana. The timber must be seasoned in order to make it light enough to float.

ring-barked (which kills the tree) and left for its leaves to wither and its trunk to season where it stands; the reason for this is that fresh-felled teak is too heavy to float to the sawmill. The tree is felled a year later, and elephants, guided by their *mahouts*, haul the logs to the nearest river, where they are lashed together and floated as rafts to the sawmill.

The problem of the broad-leaved evergreen forest, then, is not so much shortage of timber as shortage of commercially desirable species and the difficulty of harvesting them. Teak is only one example of this; others include the African tola (*Pterygopodium oxyphyllum*) and the Honduras mahogany (*Swietenia macrophylla*), both valuable for furniture-making, and the greenheart (*Ocotea rodiaei*), from Guyana, which is the best timber for piers and wharves because it resists attack by marine borers. Doubtless, with the growth both in our knowledge of tropical timber and in the industrial applications of wood, we shall eventually find uses for many more broad-leaved evergreen species. Until that time, however, we shall hardly have begun to touch the immensely rich resources of the tropical forest.

15 Fungi, Seaweeds, and Peat

Above: slime mould, *Stemonitis fusca*, in its reproductive phase; the brown, slender spore-cases, about half an inch long, are almost ready to release spores

Micro-organisms—that is, life forms that are individually too small to be seen with the naked eye—play a very important role in the lives of both plants and animals. We have already noted (page 17) the work of one group of micro-organisms, the bacteria, in the carbon and nitrogen cycles; and we have seen that bacteria and yeasts are essential to the complicated processes of fermentation by which plant sugars are converted into other organic substances.

Micro-organisms fall into one of five main groups: yeasts, bacteria, fungi or moulds, viruses, and rickettsias. Many of the most serious diseases of both plants and animals are caused by bacteria, viruses, and rickettsias. The organisms feed on the plant or animal

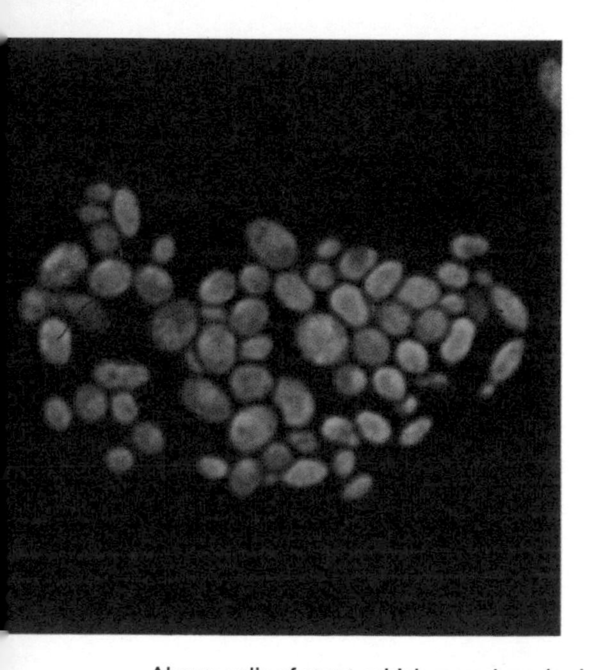

Above: cells of yeast, which reproduce (as in drawing, right) by dividing into two. Yeast cells were discovered in 1680, but it was not until 1859 that Louis Pasteur proved that they are living organisms capable of reproduction, and that they cause fermentation of sugars.

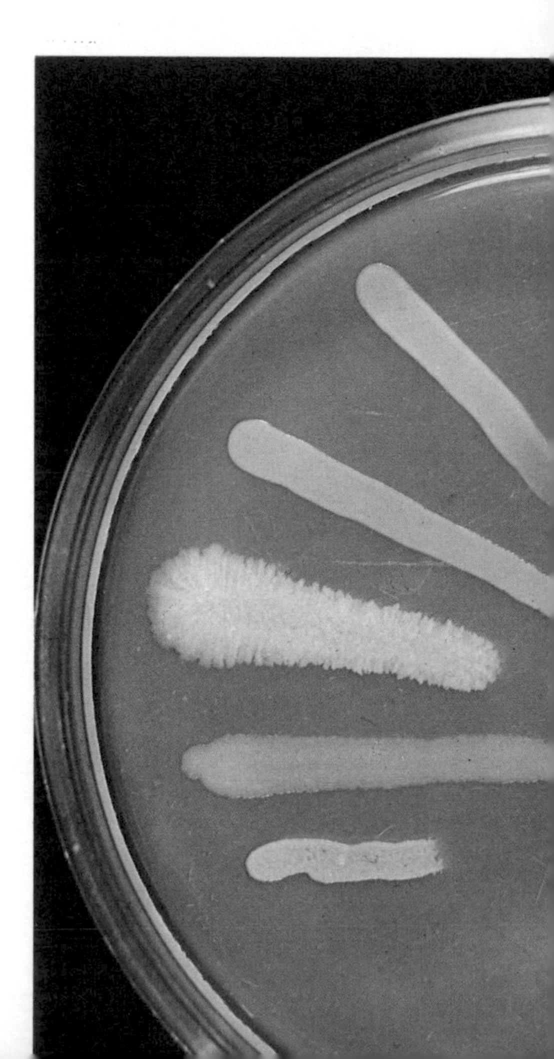

Below: the round body on the dish is penicillin, the long bodies to the left are bacteria. Note that the penicillin is inhibiting the growth of three of the bacteria (staphylococci) but not of the other species.

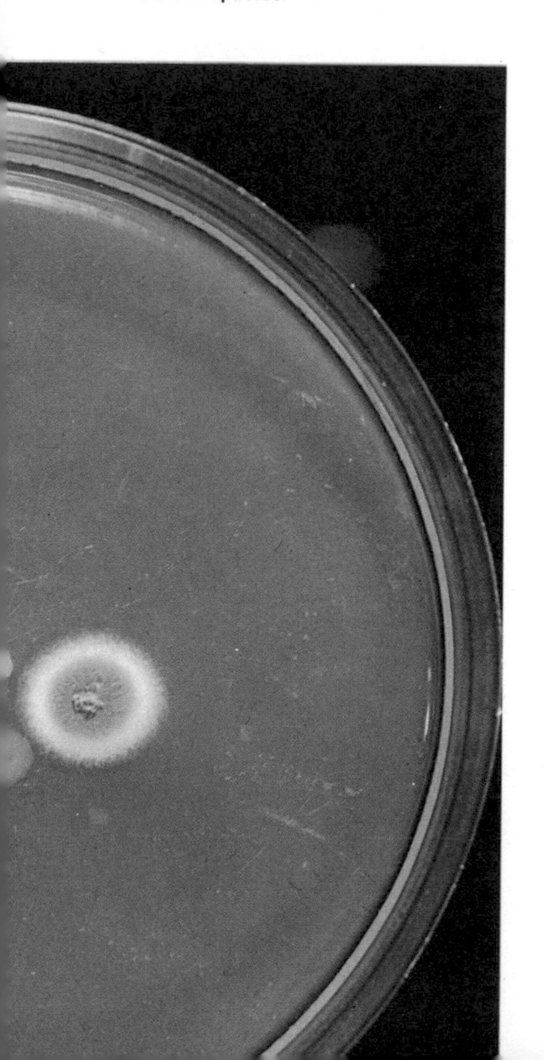

tissues they invade, and so multiply. In certain cases, the plant or animal is able to combat the infection with naturally produced substances called *antibodies*; in others, however, the micro-organisms are either too powerful or present in too great a quantity to be defeated without outside help.

Centuries ago, women in Europe used pieces of mouldy bread to dress the wounds of soldiers. In some cases, the treatment was effective—but it has been only during the last 40 years or so that we have begun to understand why. The important point is that some micro-organisms are antagonistic to others: they secrete substances that either destroy them or stop their further growth. One of the first people to recognize the potential value to medicine of this antagonism was the British bacteriologist Alexander Fleming. In 1928, while working in a laboratory at St. Mary's Hospital, London, Fleming noticed that a spore of the fungus mould *Penicillium notatum* had accidentally fallen into a culture of staphylococcus, a bacterium that causes blood poisoning, and had killed it. It was not until World War II, however, when the need for new drugs and medicines intensified, that *penicillin* (as the anti-bacterial substance in *P. notatum* was called) was isolated from its mould and produced for medical use. At about the same time the term *antibiotic* (which means "against life," a reference to their capacity to destroy other organisms) was applied to penicillin and to another substance, *streptomycin*, which was discovered in 1944. Since then, more than 20 antibiotics have been discovered and are now widely used.

In order to be medically useful an antibiotic must destroy or inhibit the growth of bacteria, viruses, or rickettsias without damaging the tissues of the host—that is, the infected plant or animal; it must, moreover, be able to reach and remain in the infected area in sufficient concentration to act effectively. *Bacteriostatic* antibiotics (those that inhibit the growth of the micro-organisms) are often just as effective as *bactericides* (which kill them): by preventing the multiplication of bacteria cells they enable the host's natural tissue defences to overcome the infection. The principal antibiotics in use today, apart from penicillin and streptomycin, are the *tetracyclines*, *erythromycin*, and *chloramphenicol*. Each has specific uses as well as some applications that overlap with the others. As a group they have greatly reduced the dangers of such major diseases as pneumonia, diphtheria, tuberculosis, meningitis, dysentery, scarlet fever, typhoid fever, and tetanus.

Above: mushroom cultivation. Spores are fed (left) an agar and nutrient mixture so that they will grow into hyphae, or spawn, which are then put into a grain medium (in bottle) ready for packing. If the spawn culture is satisfactory it is used to inoculate sterilized grain (above) to produce commercial quantities of spawn. The grain spawn is in process of "germinating" (right).

Other Useful Fungi

Fungi, like most bacteria, contain no chlorophyll and so cannot manufacture carbohydrates by photosynthesis. They live by feeding on living or dead organic material or by establishing a biological association with other living plants, such as algae, that are capable of photosynthesis. Fungi also differ from other members of the plant kingdom in having no root, stem, or leaves; they reproduce not by forming an embryo but by developing spores that grow directly into new organisms.

Many fungi are poisonous or are a nuisance to man in other ways: several, for instance, attack cultivated plants such as grains and fruits. Others, however, are extremely valuable—notably those from which antibiotics are derived. Another useful fungus is the common mushroom (*Agaricus campestris*). Like other fungi and bacteria that live on dead plants or animals, the mushroom has an important place in nature because it reduces, to forms suitable for use by other organisms, the carbon, hydrogen, nitrogen, and other elements contained in the tissues of its dead host.

The mushroom begins as a mass of spores that, when suitable nutrients are available, develop into threadlike processes called *hyphae* that spread out through the soil, breaking down the organic substances they need for food. Eventually, after a period of growth and storage of food, the hyphae (known collectively as the *mycelium*) send up a group of mushrooms. Each mushroom is essentially a *sporophore*—a container of spores—and its role in nature is to ensure the survival of the species; the mushroom head is constantly shedding vast numbers of individual spores that

Below: (1) reindeer moss; (2) *Usnea subfloridana*, a lichen containing antibiotic usnic acid, used in treatment of wounds; (3) another lichen, *Lobaria pulmonaria*, once prescribed for chest complaints owing to its resemblance to a human lung.

give rise to new hyphae. The mushroom depends for at least part of its food on nutrients contained in the dung of horses and other grazing animals. In cultivation, it can be raised only by providing the mushroom "spawn" (hyphae) with a prepared bed of fresh stable manure, which is usually mixed with peat. Since fungi do not need light for growth, they are commonly raised in sheds or cellars; the development of the crop is aided by the heat generated by natural fermentation of the manure. Each crop takes about 10 weeks to grow, after which the "spent" manure is used as a fertilizer for other plant crops.

Lichens are an example of *symbiotic* fungi—species that live in mutually beneficial association, or *symbiosis,* with other plants. A lichen comprises two organisms, a fungus and an alga. Commonly found on bare rocks, old stone walls, and other places where the soil is too thin to support higher plants, lichens often consist of a crust-like fungus enclosing a large number of single-celled algae. For the fungus, the association has the advantage that the algae are capable of photosynthesis, so that the lichen plant as a whole does not have to rely on the meagre supply of nutrients available from the soil. One species of lichen, reindeer moss (*Cladonia rangiferina*), flourishes in the arctic tundra, where the thin soil is frozen for much of the year and supports little other plant life. This lowly plant is central to the existence of the Lapp peoples of northern Scandinavia. The lichen forms the principal diet of the reindeer in this region; in turn, the reindeer provide the Lapps with meat and milk, skins for clothing and tents, bone and horn for tools, and also act as draught and pack animals.

Mushroom growth (right) over a period of seven days. Photos were taken at 24-hour intervals. Spawn had been placed in compost overlain by chalk-peat mixture 14 days before first picture was taken.

Seaweeds

Most seaweeds belong to the large plant group, Algae. As we have mentioned, algae are capable of photosynthesis; thus, seaweeds in this group are confined to water shallow enough for the sun's rays to penetrate. Although seaweeds contain a starch-like substance called *laminarin* and a sugar-like substance called *mannitol*, they are not readily digested by man, though a few species are eaten by people in certain regions. One, called carrageen or Irish moss (*Chondrus crispus*), is gathered on the rocky coasts of Portugal, Brittany, and Nova Scotia, and is used mainly in medicine and in making blancmange. Agar-agar, derived from *Gracilaria confervoides*, a seaweed found on the shores of Malaysia, Indonesia, and Japan, provides a jelly, which is used medicinally and as a culture medium for bacteria and fungi. Cattle and other ruminants can, however, digest many species of seaweed, and on the rocky shores of western Scotland, highland cattle and hill sheep often feed on seaweed stranded on the shore after storms.

In many countries seaweed is harvested for manure. It contains all the mineral nutrients and many of the trace elements needed for plant growth. In the Channel Islands, for example, the seaweed, or *vraic*, is regularly

Above: gathering the seaweed *Porphyra tenera* in Japan. During winter spores of the seaweed become attached to bundles of brushwood driven into the mud. The mature seaweed is torn from the bundles, dried over a fire, and used to flavour soups.

Right: *Laminaria digitata*, a seaweed common on British shores. A rich source of potassium, it is sometimes used as a component of fertilizers.

Above: harvesting *Ascophyllum nodosum* at Froya, Norway. This seaweed is a source of alginates.

gathered for fertilizing the soil for the potato crop.

At one time kelp-burning was an important industry along the shores of western Europe. Kelp is the ash of seaweed, and to make it the weed had to be gathered from the shore—or from deeper water using boats—dried on the beaches, and then burnt in stone-built kilns. Kelp is rich in soda, potash, and iodine, and until the late 19th century was a leading source of these minerals for the chemical industry. Since then other sources for these materials have been found and kelp-burning has died out.

Seaweed remains important to the chemical industry, however, because it is the main source of a group of chemicals called *alginates*, with several useful properties. Alginates are used in the preparation of emulsions, and as thickening agents in certain prepared foods—ice-cream, soups, sauces, and others. Certain alginates can be made into soluble textile fibres for special purposes. This property of solubility is also important in medicine. Alginate pads containing antibiotics are often used for dressing damaged tissues, as in serious burns. When the antibiotic has done its work the pad dissolves and so eliminates the need to remove it (often causing pain and further tissue damage) before applying a fresh dressing.

Agar is used as a medium for testing antibiotics, which have been placed in the eight small trays. The device enables the chemist to compare the resistance of a disease organism to each antibiotic.

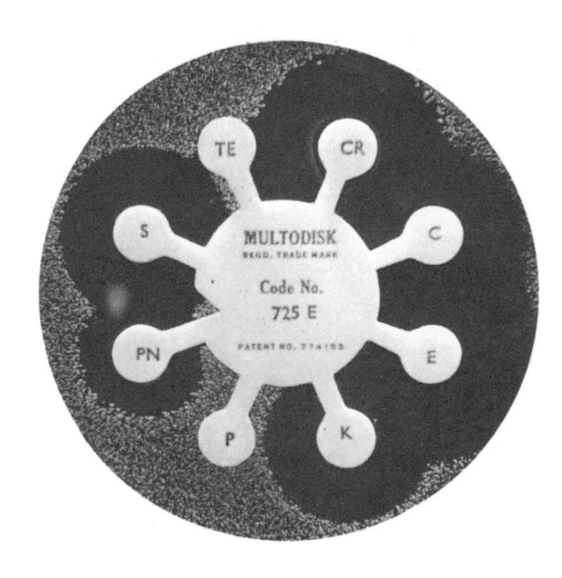

Peat

When they die, most plants decay rapidly, because their leaves, stems, and roots are attacked by bacteria and fungi. But if the plants happen to be in a watery environment, such as a swamp or a boggy soil, this process of decomposition cannot proceed normally, for the fungi and bacteria that break down the plant tissues need oxygen as well as moisture in order to live. On marshlands, therefore, an ever-thickening layer of plant tissues, only partially broken down, may gradually develop. All the world's coal originated from the remains of vast swamp forests, whose trees died, partially decomposed, became compacted under enormous pressure, and eventually fossilized.

Peat bogs, which are found in the colder countries such as Canada, northern Siberia, Scandinavia, and the British Isles, consist of deep layers of decayed plant tissue. They hold water like a sponge, and so the plant material steadily accumulates, since it cannot entirely decompose. The plants that produce it can grow only on the top of the bog, in the thin surface layer where their roots get enough air. Many are mosses, especially of the genus *Sphagnum*, but heather and certain grasses also form peat. As their leaves and stems die and fall, the peat becomes deeper and the living plants rise higher. The process may continue for thousands of years, forming a peat bed many feet thick.

In those northern lands that are too cold for trees to flourish, or where—as in much of Ireland and Scotland—the forest cover has been destroyed by man, peat has great value as a fuel. Peat beds are a form of property, owned by a particular farm or village. The peat is full of water, and has to be dried before it will burn, so it is always prepared in the summer months when the weather is warm and dry. Using sharp spades, the men cut the soft turf into oblong pieces that are stacked in open heaps to catch the wind. When winter approaches, the peat is dry enough and light enough to be carted to the homestead. It burns slowly, without a flame, and a peat fire, once lit, is seldom allowed to go out. Many northern countryfolk rely entirely on peat for fuel. Peat is also used in horticulture, partly for the nutrients that it contains, but mainly because it keeps soil in good condition and holds moisture available during dry weather.

In Eire, the vast resources of peat are exploited by a government authority. Much of the peat, which is collected mainly in the counties of Kildare, Offaly, Tipperary, and Meath, is being used as a fuel in the generation of electricity.

Cutting (above) and transporting (below) peat in County Galway, Eire. The Irish have used peat as a fuel for centuries.

Right: this massive machine cuts peat into small bricks, which are used to fire the boilers of Allenwood, Eire, power station, in the background.

16 Commercial Flower Growing

Our love of beautiful things probably goes back to man's beginnings. Even hunter-gatherers, whose lives seem to us to have consisted of little else but an unrelenting search for food, found time to make astonishingly fine paintings on the walls of caves, as at Lascaux and Altamira. Doubtless flowers were amongst the first objects that man considered beautiful. But the great horticultural industry that caters to our pleasure in flowers is a product of urban civilization, for it depends mainly upon large numbers of city dwellers (and also on commercial and industrial firms) who do not own gardens but like to have cut flowers around the house or office. Flowers and flowering plants are perishable. Until the coming of the railways, which enabled plants to be sent rapidly from one part of the country to another, the products of flower markets were sold only locally. With the development of air transport the flower industry has expanded still further; nowadays, many flower crops are raised in warm regions many hundreds of miles from the countries where their main customers live. It is possible to send flowers halfway around the world in less than a day; it is an expensive business, but many people are prepared to pay high prices for exotic blooms. Alternatively, as we shall see, the local climate can be altered by using glass-houses.

There are three stages in their life-cycle at which flowering plants are commonly sold. The first is as planting material—that is, seeds, bulbs, or corms, which the customer buys in the "resting" state to grow in gardens, pots, or bowls. This is a relatively unimportant market, for it is seasonal and many people lack the space or the time to raise flowers in this way. The second market is for partly-grown plants in pots, bowls, or baskets, which the buyer can bring into flower with little trouble. This, too, is seasonal and limited in extent.

The third and most important market is for cut flowers just coming into bloom. Nowadays people

The commercial flower industry offers the customer everything from cut flowers (such as the chrysanthemum, left) to an immense variety of seeds, bulbs, and pot plants (above).

expect to have blossoms all the year round, whatever the weather may be, and as flowers stay fresh for only a few days, many people buy a fresh supply every week. The size of this market owes a great deal to commercial and industrial firms who have long-term contracts with growers for flowers to decorate their offices.

Some growers aim to meet all three demands. For example a bulb-grower may sell part of his crop as "dry" bulbs for other people to grow in bowls or to plant in pots, another part as growing bulbs already planted in pots, and yet another part as cut flowers. Other growers concentrate all their efforts on a single kind of crop: for example, they may grow nothing but chrysanthemums, which they persuade to bloom the whole year round. Others breed new varieties, or raise seed and other planting material for the commercial growers who sell flowers to the public.

The Life-Cycle of the Flowering Plant

We saw, earlier, when discussing plants grown for food, that man can do little to alter the basic life-cycle of a plant. If by nature it grows as an annual from seed, a perennial from a bulb, or as a woody shrub, he must accept this as his starting point. The value of flowers is so high, however, that some growers spend a good deal of money and effort on manipulating a plant's life-cycle to suit the purchaser's needs.

Only a few species of flowers are grown on a big scale for sale, though there are hundreds of others that could be used. The growers seek plants that are easily handled when cultivated on a large scale. Many choice garden plants and shrubs need too much care, take too long to flower, or show too little variation of form and colour for their cultivation to prove commercially worth while. The desirable qualities of a commercial flowering plant suitable for modern markets are these:

(a) It must be easy to propagate and to grow to the flowering stage;
(b) It must respond to treatments that give it a long flowering season, preferably the whole year round;
(c) It must show ample variation so that, by skilful breeding, many varieties with different forms and colours can be obtained;
(d) It must stand up to transportation and handling, and stay in flower for several days at least, giving a bold display on a tall stem.

We shall see how these needs are met by a small selection of flowers, but first let us consider how man can alter the *environment* in which they are grown.

The massive scale of operations of many nursery firms is exemplified (above) by this glass-house, half an acre in area and containing 40,000 plants (poinsettias).

Controlling the Plant's Environment

The most effective way to control the environment of a plant in a number of important ways is to use a glass-house. Glass-houses were first built so that tender tropical plants could be grown in temperate climates, but today they have many uses in commercial horticulture. The glass keeps out the cold of the surrounding air, yet holds in the warmth of the sun's rays, and artificial heat is usually provided as well. This means that the temperature, one of the major features of the environment, is controlled: the grower can give his plants warm, summer temperatures in winter.

The length of daylight, called the *photoperiod*, also affects the growth of certain plants to a marked degree. Out-of-doors, the grower has no control over day length, but in a glass-house he can make the winter day artificially long by using electric light, or the summer day artificially short by using blinds.

Since no rain reaches his glass-house plants, he has full control over their water supply. He can give as much, or as little, water as he thinks best for growth at any stage. By using sprays and ventilators, he can also control the moisture content, or *relative humidity*, of the air.

All the soil used in a glass-house, is brought in from outside, and usually it serves for only one crop before it is removed for further treatment. The grower can therefore choose his soil and his fertilizers with certainty, and he can also sterilize the soil to keep it free from pests and diseases.

Below: in this widely used method of heating glass-houses, warm air is forced along the flexible plastic pipes by a powerful fan heater.

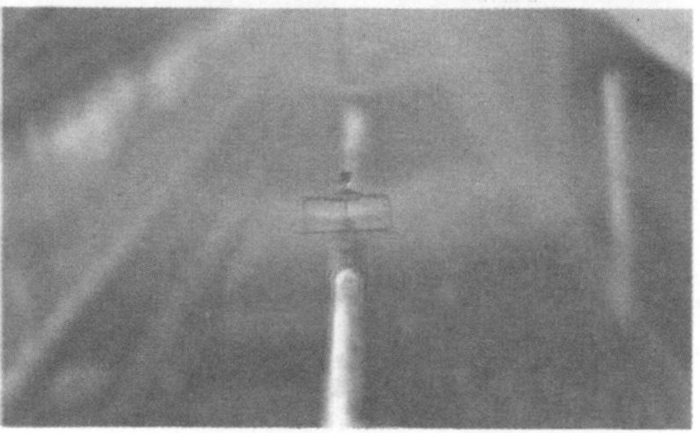

The two principal methods of watering glass-house plants: drip feed (upper picture) and mist spray (lower picture); the spray mechanism switches on and off in response to humidity variations in the glass-house.

Below: artificial lighting is used not only to supplement sunlight during the winter months but also, in some cases, to increase the length of day, or photoperiod, at other times of the year.

The Stock, an Annual Cut Flower

The wild stock (genus *Matthiola*) grows in Mediterranean lands as a short-lived plant that springs up from seed in the autumn, when both warmth and moisture are available. Although it is called an annual, it is really an *ephemeral*: its full life-cycle is much shorter than a whole year, for it can ripen its seeds within three months of starting growth. This means that growers can produce flowers the whole year round under glass by sowing seed at intervals of a few weeks. One well-known strain will open its flowers ten weeks after sowing, another one within only seven weeks. Because in nature the stock is accustomed to starting growth in autumn, when the days are getting shorter, it makes a good source of winter flowers.

The breeding of stocks begins with the selection of the best individual plants, which alone are allowed to bear seed. These may, for instance, include strains that bear double flowers with extra petals, and those that show especially bright colours. The selected strains are grown far apart from poorer kinds, so that they receive pollen only from other good bloomers: a bed of red stocks will be set a mile away from a bed of blue stocks. Seed growers always select and re-select their breeding material, as otherwise it would deteriorate, and only poor flowers would result.

One attraction of the stock as a cut flower is its succession of blooms; each stalk carries a series of buds, and those at the base open first. It is popular with gardeners, too, because it can easily be transplanted at the seedling stage. So we find that there are three groups of growers who make money from this easily grown, short-lived plant. There are the seed-farmers, the nurserymen who sell seedlings to gardeners, and the commercial flower-growers, owning glass-houses, who can send stocks to market the whole year round, though they make their main effort in midwinter, around Christmas time.

The Tulip, a Perennial Bulb

Many kinds of tulip (genus *Tulipa*) grow wild in Turkey and neighbouring lands. Their seeds produce seedlings that slowly develop into bulbs, which are clusters of swollen scale leaves set well below ground. The tulip stores water and food within the bulb, which remains dormant through the winter. Then, in the spring, it sends up a flowering shoot that is nourished by the bulb's food reserves. But a bulb that has grown from seed will not flower before its third or fourth year.

In practice, the only people who regularly raise tulips from seeds are the plant breeders, who develop new

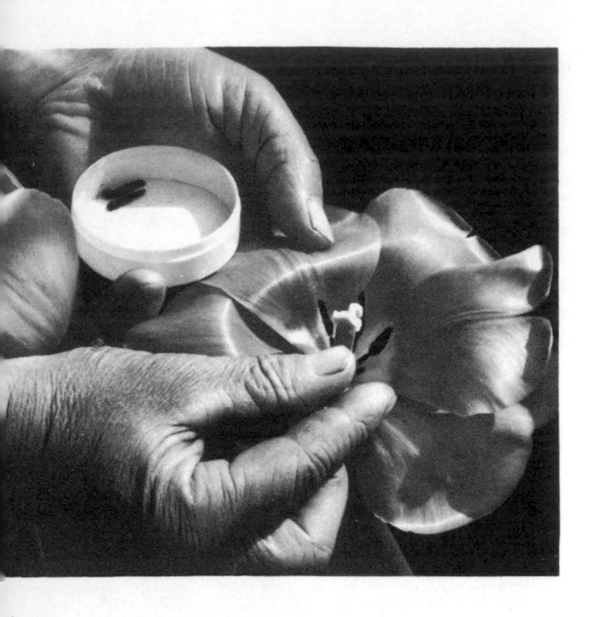

Above: removing stamens from a tulip flower, which will then be fertilized by pollen from another variety to produce hybrid seeds and bulbs. If the hybrid shows promise, it can be increased by detaching the bulbils, or offsets (below), that develop on each bulb.

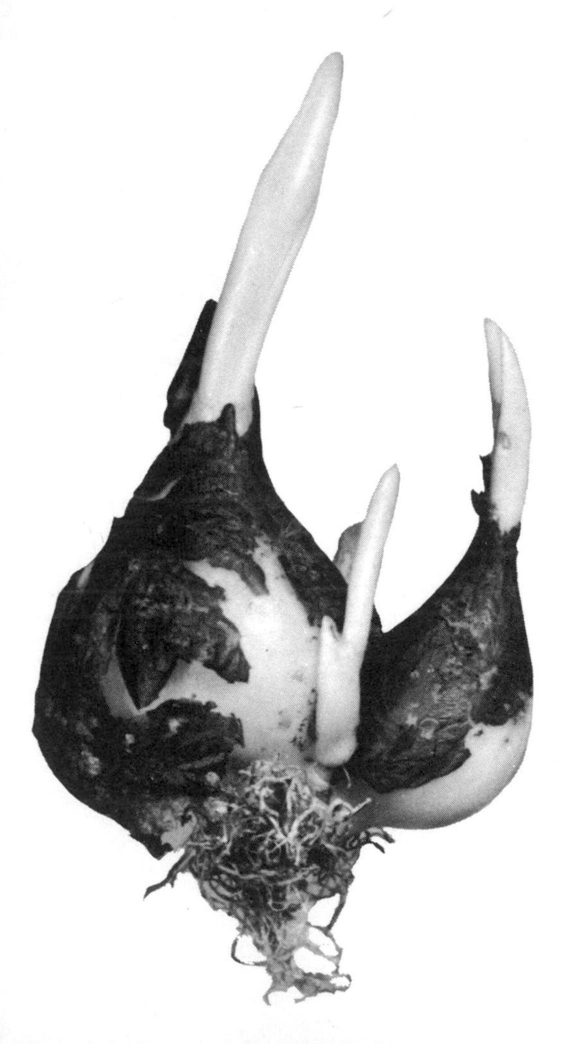

varieties by crossing one kind of tulip with another. To do this, they will select one kind as the female parent, and remove the male stamens from its flower, so that it cannot pollinate its own seeds. Then they select another flower as the male parent, and using some simple device, such as a syringe or even the fur on a rabbit's paw, transfer pollen from this flower to the female parent. After the seed has ripened, they sow it carefully in the following spring, and raise a crop of hybrid bulbs.

Four years later, when these bulbs come into flower, the breeders can tell for the first time whether all this work and waiting has been worth while. If they have crossed, say, a yellow tulip with a red one, they will get a group of flowers of various colours from yellow, through orange, to red. Most will be of little value, but one or two may stand out as desirable new kinds. Success is not assured yet, however, because these attractive hybrids may prove very susceptible to pests or diseases. Scientists are still trying to find out just what factors make certain strains of plants resist disease better than others. The breeder must carry out practical tests to discover how each new hybrid behaves when exposed to infection or to conditions that favour attack. Many hybrids have as their parents two separate species of plant, each chosen to contribute some useful factor. For example, a species with small flowers, but known to resist disease, may be crossed with a large-flowered kind, in the hope of combining big flowers with disease resistance.

Once a desirable tulip has been found, it is a simple matter to increase it, though this takes time. Tulip bulbs produce *bulbils*, or offsets, at the base of the scales that make up the main bulb. The bulbils develop roots and grow, in a year or two, into full-sized bulbs. Because this method of increase is *vegetative*, and does not involve the cross-pollination needed to produce seed, each new bulb bears flowers exactly like those of its parent. So the grower always increases the working stock by means of bulbils.

Vegetative reproduction, by bulbils, offsets, cuttings, or grafts, is very important in commercial flower-growing. Each single strain is called a *clone*, or a *cultivar*—that is, a variety that has been developed in cultivation—and it is given a name that is registered with a horticultural society. People who buy tulips under the cultivar name know that every bulb in a group will produce a flower exactly like its neighbour. One great advantage of vegetative propagation is that many varieties of the same kind of plant can be grown side by side with no risk of mixing.

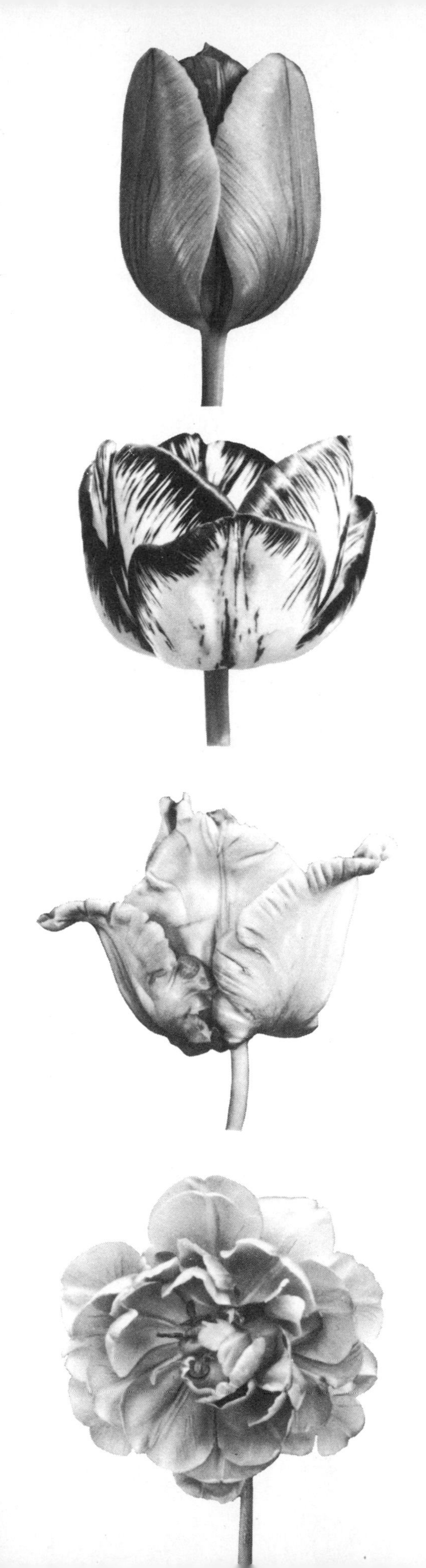

The tulip bulbs that we buy in the shops are raised on a big scale by growers who plant small bulbs and bulbils in long rows in fertile, well-cultivated, and well-drained soil. This is done in autumn, and next spring the bulbs send up shoots and start to flower. At this stage the growers "top" the plants by cutting off all the flowers with a sharp knife. This stops the plant using its carefully stored supplies of nourishment for the formation of seeds, and so, when the leaves and the flower stalk fade, all the plant food goes to the bulb.

The bulb, in consequence, grows exceptionally large and at the same time forms a flower bud that will bloom the following spring. In the autumn the grower digs up all the bulbs, cleans them, dries them, and sorts them. He sells the big ones to gardeners, and replants the small ones, and also the new bulbils that have formed, to provide his next crop.

Enormous quantities of tulip bulbs are sold every autumn to home gardeners or to professionals who look after public parks. Some are raised as pot plants under glass, and sold to housewives just as they come into flower; others are used to provide cut flowers. The latter are gathered just as their buds change colour, from green to one of the many beautiful tulip shades that range through yellows, reds, and blues, with many patterns and colour blends.

Tulips are available from Christmas until the end of June. This long season is due to the number of different varieties that are available. If one buys a range of tulip cultivars and plants them all on the same day in mid-October, some will bloom early in March, others in April and May, and still others as late as June. This growth period varies from one kind to another, but is constant for each named kind.

Many years ago bulb-growers in Holland learned how to prepare tulip bulbs for early flowering by the use of temperature control. They lift short-season bulbs out of the ground in midsummer. First the bulbs are stored at a fairly warm temperature, about 65°F, until they have formed (within the dry bulb) the actual flowers that will appear later. Next, the bulb-growers chill the bulbs to 48°F and keep them at this unnaturally low temprature for some weeks through August and September. This stops growth just as the natural cold of winter would do. When the prepared bulbs are planted in October, in warm soil within a glass-house, they start to grow at once and, provided a suitable temperature is maintained, they open their flowers at Christmas, instead of in March. In effect, they have been through a "false winter," three months before the real one.

Left: the tulip at the top is a common variety known as Sonata. The others are hybrids and show how both the shape and colour of tulips can be altered radically by cross-breeding.

Raising Hyacinths

The tulip, and many other bulbs (such as daffodils or narcissi), are easily increased by bulbils that form naturally. But the bulb of the hyacinth (genus *Hyacinthus*) rarely yields such offshoots, for it is simply a storage organ. Wild hyacinths grow only from seed. Growers increase flowering hyacinths in a curious way. They take a large bulb and scoop out its base with a sharp knife. This wounds all the leafy scales and kills the main flower bud at the centre. The damaged bulb is then stored in a dark chamber at carefully controlled temperature and humidity. Soon each damaged scale slowly develops a tiny bulbil, and after this treatment has gone on for several months, from August to November, the parent bulb, with its bulbils still attached, is planted in outdoor beds. In its early years each bulbil grows only leaves, but later it produces flower buds. The growers cut off the flowers before they can bear seeds, and eventually—after five years of growth—a fine top-size bulb, holding a strong flower spike, develops.

Chrysanthemums

Just as the Dutch have perfected bulbs such as tulips, daffodils, and hyacinths, so the Japanese have developed a beautiful and adaptable herbaceous perennial plant, the chrysanthemum (genus *Chrysanthemum*). Each new variety of chrysanthemum is raised from seed, but later on it forms a rhizome that is easily divided. Once a clump of chrysanthemums has flowered, the young shoots that arise at the base of the flowering stalks are separated as rooted cuttings. All the many named varieties are rapidly increased in this way.

No plant has rewarded the breeders so well for their efforts. One can buy varieties of chrysanthemums with single or double flowers, in a great range of colours, and with short or tall stems. Commercial growers prefer the long-stalked kinds as cut flowers, and the dwarf varieties for pot plants. Showy kinds with incurved petals, displaying shades of silver or gold on their outer sides, and rich reds, browns, or yellows within, fetch high prices as single blooms; in contrast, the smaller-flowered varieties are sold in bunches. Many modern kinds are hardy enough to bear rich blossoms out-of-doors. One great merit of the chrysanthemum is that it flowers in autumn, when other flowers are getting scarce. By a skilful handling of glass-house temperatures, growers have learnt how to produce fine flowers under glass right up to Christmas time.

Hyacinth bulb (top) has its underside removed with a knife. After a few months under controlled conditions (below), the "mother" bulb develops between 20 and 50 tiny bulbils (centre), depending on its size. The bulb, still with its bulbils attached, is then planted out. The bulbils grow steadily each year and, after five years' growth, have developed into mature adult bulbs.

Horticultural Research

Farmers have always set aside their best plants to yield seed for future crops, but the planned breeding of new varieties of plants originated with the flower-growers. Gardeners who wanted new varieties of tulip, for example, were the first people to make deliberate crosses between different plants. Now this technique of *hybridization* is used regularly by farmers who want to produce a new kind of wheat or potato.

The flower-grower, especially if he raises plants under glass, needs to know a great deal about the limited amount of soil he uses. Much that we know about soil chemistry, and the part played by water in plant growth, has come from the work done by horticultural research stations. Working on a fairly small scale, the flower-grower can control the environment of his crop in a way that the farmer, with far bigger fields, can never hope to do. But though the farmer cannot cover his land with a glass roof, he can make profitable use of the flower-growers' scientific discoveries. As we have seen, flower-growers have discovered that there are some *short-season* strains of common plants, such as a stock flower, that bloom after seven weeks of growth instead of the usual ten weeks. Farmers in the far northern countries, such as Siberia and Canada, need a short-season wheat to make use of their brief frost-free summer. By applying the same principle—that rapid growth and seed-ripening are inherited features of particular strains of plant—they have been encouraged to seek and find suitable wheat varieties.

The flower-growers' discovery of the importance of photoperiod is being applied by two groups of people who raise very different kinds of useful plants. Foresters now know that it is inadvisable to use seeds from trees that grow in a region of long summer days and plant them in a region of short summer days. Spruces from Alaska will not grow happily in England because, in midsummer, they miss the midnight sun. Pastoral farmers found similar problems when they moved a strain of grass from Wales, where days are short in winter and long in summer, to the cool tropical climate of the Kenya Highlands, where day-length is the same the whole year round.

When man first began to cultivate plants for their flowers, he asked himself the same questions as he did about his food plants. How do they grow, and what causes them to vary in size, and form, and colour? By seeking answers to these questions he has increased his knowledge of plant life as a whole, and his mastery over his means of existence.

Growing habits of flowers are studied in these cabinets (above), at Wye College, Kent, in which temperature, photoperiod, and light intensity are controlled. The plant below (*Kalanchoe blossfeldiana*) flowers only if exposed to short days and long nights. The example on the right received this treatment continuously during its flowering phase; that on the left for only 96 hours.

Index

Numbers in *italics* refer to illustrations and captions to illustrations.

Acknowledgments

Key to picture position: (T) top, (C) centre, (B) bottom, and combinations, for example (TL) top left, or (CR) centre right

6 The Industrial Forestry Association
10 & 11 (L) Photos Gene Cox, by courtesy of South West Optical Instruments Ltd
11 (B) Reproduced by kind permission of the Metropolitan Water Board, London
13 (T) Photo K. L. Shumway
22 (BL) Photo Gene Cox, by courtesy of South West Optical Instruments Ltd
24–5 (TC) Photo J. Arthur Herrick, Kent State University
26 British Museum (Natural History)
29 (B) Uni-Dia-Verlag, Stuttgart
30–1 British Museum
33 British Museum
34–5 British Museum (redrawn by Nina G. Davies)
36 (B) British Museum
37 Lockwood Survey Corp.
40 (T) & 41 (T) Photos D. Mackney, Soil Survey of England and Wales
44 (T) Map by FAO
46 (B) & 47 Fisons Limited
48 (T & B) British Museum
50 (T) Plant Protection Ltd (Imperial Chemical Industries)
51 (T) A Shell photo
(C) Paul Popper Ltd
52 British Museum
53 Photo Paul Almasy
56–7 (B) Photo Ben Darby
58 (T) British Museum
(BL) MS. Canon. Liturg. 99, f.16, Bodleian Library, Oxford
(BR) Photo Paul Almasy
63 The Distillers Company Limited
65–6 British Museum
67 (BL) Ewing Galloway
(BL) Ewing Galloway
(BR) Photo Paul Almasy
68 (B) Paul Popper Ltd
69 (B) Photo Paul Almasy
70–1 Victoria & Albert Museum, London
72 (T) J. Allan Cash, F.I.B.P., F.R.P.S.
(C & B) Paul Popper Ltd
73 (T & B) Paul Popper Ltd
74 Photos Paul Almasy
75 (T) FAO photo
76 MS. Ashmole 1431, Bodleian Library, Oxford
78 (T) Museum of English Rural Life, The University of Reading
(B) John Topham Ltd, Sidcup, Kent
81 (BR) Soil Conservation Service, U.S. Department of Agriculture
83 (B) Fisons Limited
84 (T) *Farmers Weekly* photo
86 (B) & 87 (T) John Topham Ltd, Sidcup, Kent
87 (B) Standard Oil Company, New York
89 (T) British Museum
90–3 Photos supplied by courtesy of Tate & Lyle Limited
94 (T) Photo Kenneth I. Oldroyd, Wisbech
98 (R) & 99 (B) Photos Kenneth I. Oldroyd, Wisbech
100 (T) Photo Kenneth I. Oldroyd, Wisbech
(Lower 3) British Sugar Corporation
102 Shell photo
103 (TL & TR) Shell photos
(CL) Photo *World Crops*
(BL & BR) South Pacific Commission, Sydney
104 (R) Musée de l'Homme, Paris
105 (T, B, & C) Photos by courtesy of the Potato Marketing Board
106 Photos Kenneth I. Oldroyd, Wisbech
107 (T) Mansell Collection
(B) Imperial Chemical Industries Ltd
109 British Museum
111 Parkinson: *Paradisi in Sole Paradisus Terrestris* (1629) London Library
113 (T) Photo H. Smith, Westcliff-on-Sea
115 (TL & TR) Photos Kenneth I. Oldroyd, Wisbech
(B) Ministry of Agriculture, Fisheries and Food
116 British Museum (Natural History)
118 (T) British Museum (Natural History)
(B) Fropax Eskimo Frood Ltd
119 (T) British Museum (Natural History)
(B) Photo Paul Almasy
122 (T) Glasshouse Crops Research Institute, Littlehampton
(B) Fox Photos Ltd
123 (T) Koninklijk Instituut voor de Tropen, Amsterdam
(B) Horniman Museum, London
124–5 (B) British Museum (Natural History)
126–7 Photos courtesy Fyffes Bananas
128 British Museum (Natural History)
129 Photos courtesy Dole Company, Honolulu
130 British Museum (redrawn by Nina G. Davies)
131 British Museum
134 (T) Royal Botanic Library, Kew
134 (B) & 135 Duhamel de Monceau: *Les Arbres Fruitiers* (1835). British Museum
136 (T) J. Allan Cash, F.I.B.P., F.R.P.S.
(B) Photo Paul Almasy
137 (T) J. Allan Cash, F.I.B.P., F.R.P.S.
(B) Photo *World Crops*
138–9 East Malling Research Centre
140 (B) Department of Agricultural Engineering, University of California
142 British Museum
143 (T) Picturepoint, London
(B) Jewish Museum, London
144 (T) J. Allan Cash, F.I.B.P., F.R.P.S.
(B) Photo Paul Almasy
146 (B) Photo Bradley Smith, Gemini Smith Inc.
147 Photos Paul Almasy
148 German Tourist Information Bureau, London
149 (T & BR) Picturepoint, London
(BL) MS. Corp. Christi Coll. 285, courtesy Christ College, Oxford
150–1 (B) French Government Tourist Office
154 (B) Bibliothèque Nationale, Paris
155 (B) Photo Ceylon Tea Centre, London
156 (T) Photo Ceylon Tea Centre, London
(B) Photo courtesy Brooke Bond & Co. Ltd
157 (T) Victoria & Albert Museum, London
158 (TR) J. Allan Cash, F.I.B.P., F.R.P.S.
158 (B) & 159 Photos courtesy Lyons Coffee
160 (TR & BR) Paul Popper Ltd
161 (TL & TR) Paul Popper Ltd
(BR) Courtesy Cadbury Brothers Limited, Bournville
162 (TR) Fox Photos Ltd
(BL & BR) Courtesy Wigan Richardson, Hop Merchants
163 Photos J. Allan Cash, F.I.B.P., F.R.P.S.
164 (T) Jacques d'Aléchamps: *Historia Generalis Plantarum*, Vol. I (1587)
(B) British Museum
166 (B) British Museum
167 (T) Camera Press/photo Desmond O'Neill
(B) Unilever Ltd
168 (TL) Appollot Photographies, Grasse
(TR) Bulgarian Chamber of Commerce, Sofia
168–9 (B) Proprietary Perfumes Ltd, Ashford, Kent
169 Photos Proprietary Perfumes Ltd, Ashford, Kent
170 (BR) By courtesy of The Imperial Tobacco Company (of Great Britain and Ireland), Limited
171 (T) Photo courtesy Gallaher Limited
(B) Photos W. D. & H. O. Wills
172 (BL) British Museum (Natural History)
(CB) Myddleton House Drug Garden, School of Pharmacy, University of London
173 (T) William Ransom & Son Ltd, Hitchin
(BL) Pharmaceutical Society of Great Britain
174 (B) & 175 U.S. Department of Agriculture
176 William Withering: *An Account of the Foxglove and some of its medical uses* (1785)
177 (T & BL) United Nations photos
(BR) Photo Paul Almasy
178 U.S. Department of Agriculture
179 (B) William Ransom & Son Ltd, Hitchin
180 Photos Harald Schultz
181 Photos Koninklijk Instituut voor de Tropen, Amsterdam
182 (BL) Unilever Ltd
(BR) Radio Times Hulton Picture Library
183 (BL) Unilever Ltd
184–5 (TC) Photo Paul Almasy
185 (BL) Camera Press/photo Hans Silvester
186 (T) Camera Press
187 Paul Popper Ltd
188 Photos courtesy Johnson Wax
189 Forest Service, U.S. Department of Agriculture
190 (T) Sudan Embassy, London
(B) Camera Press/ photo Alistair Matheson
191 (T) B. J. Lafiteau: *Moeurs des Sauvages Americains* (1724)
(BL & BR) Photos Canada House
192 (C & B) Photos courtesy Dunlop Company Ltd
193–4 Photos Malayan Rubber Fund Board
195 Photos courtesy Dunlop Company Ltd
196 (T) British Museum
197 Graffham Weavers
198 Photos Forestry Commission, British Crown Copyright Reserved
199 (T) Messrs. Richard Hodgson & Sons Ltd, Beverley
(B) J. Allan Cash, F.I.B.P., F.R.P.S.
200–1 (T) Museo Egizio, Turin/photo G. Rampazzi
(B) Courtaulds Ltd
203 (T & C) Photos by courtesy of The Textile Council
204 (T) Photo by courtesy of The Textile Council
205 Photos John H. A. Jewell, M.B., F.R.C.S.I.
206 (BL) Uni-Dia-Verlag, Stuttgart
(BR) J. Allan Cash, F.I.B.P., F.R.P.S.
207 (T) J. Allan Cash, F.I.B.P., F.R.P.S.
(B) The Irish Linen Guild, Belfast
208 (BL & BR) J. Allan Cash, F.I.B.P., F.R.P.S.
209 (B) Photo *World Crops*
210 (T & BL) Camera Press
211 (T) Camera Press
(B) J. Allan Cash, F.I.B.P., F.R.P.S.
212 Photos by Gavin Maxwell, reproduced by permission of his agents, Peter Janson-Smith Ltd
213 (T) Photo Koninklijk Instituut voor de Tropen, Amsterdam
(B) British Museum
214 (T) A Shell photo
(B) J. Allan Cash, F.I.B.P., F.R.P.S.
215 The Cork Manufacturing Co.
216–7 (T) The Bowater Organisation
(B) British Museum
218–9 Photos Forestry Commission, British Crown Copyright Reserved
220 Microphotographs Timber Research and Development Association, High Wycombe
221 (TL) Photo courtesy Rentokil Laboratories Ltd
(TC) A Shell photo
(TR) British Museum (Natural History)
(B) Photos The Plywood Manufacturers Association of British Columbia
222 (BL) J. Allan Cash, F.I.B.P., F.R.P.S.
(BR) The Bowater Organisation
223 (TL) British Museum
(TR) Mansell Collection
(B) Photo courtesy Société de la Viscose Suisse, Emmenbrücke
224 (T) Forest Service, U.S. Department of Agriculture
(C & B) The Bowater Organisation
225 The Bowater Organisation
226 (T) The Bowater Organisation
(B) J. Allan Cash, F.I.B.P., F.R.P.S.
227 (T) Forestry Commission, British Crown Copyright Reserved
230–1 (B) Photos J. Allan Cash, F.I.B.P., F.R.P.S.
231 (T) Photo United Africa Co. (Timber) Ltd
232 (BL) Photo Gene Cox, by courtesy of South West Optical Instruments Ltd
232–3 (T) Ian K. Ross, University of California
(B) Beecham Laboratory, Betchworth
234 (CR) Photo Dr. Teuvo Ahti, University of Helsinki
(BL & BR) Photos K. L. Alvin
236 (T) Victoria & Albert Museum, London
237 (T) Norwegian Institute of Seaweed Research, NTH, Trondheim
(B) Oxoid Ltd
238 (T) Paul Popper Ltd
(B) Camera Press
239 J. Allan Cash, F.I.B.P., F.R.P.S.
244 (T) Photo Bulb Information Centre, London
245 Photos J. E. Downward, F.I.B.P.
246 Photos Bulb Information Centre, London

Front and back end papers Duhamel de Monceau: *Les Arbres Fruitiers* (1835) British Museum

PHOTOGRAPHERS' CREDITS

The following photographs have been taken specially for this book by:

R. Benfield: 172 (BC), 172(BR), 173(B)

Ken Coton: 60(B), 143(B), 183(BR), 185(BR), 186(B), 190(C), 210(BR), 215

Geoffrey Drury: 20(B), 22(T), 24–5(B), 46(T), 49, 57(T), 77, 97, 101, 112–3, 120–1, 133, 138–9, 140(B), 141, 145, 165, 173, 222(T), 234(T), 235(TL), 235(R), 236(B), 241, 242–3, 247

John Freeman: 59(T), 117

Behram Kapadia: 108, 114, 116, 118(T), 123(B), 124(B), 125(B), 128, 179(T)

John Webb: 2–3, 34–5, 52, 65, 125(T), 130, 131, 134–5, 142, 152–3, 157, 196(B), 197(R), 213(B), 228(R), 229(R), 236(T), 240, 254–5

ARTISTS' CREDITS

Brian Lee: 183(T), 184(B)

Jill Mackley: 28(TR), 28(BR), 54(TR), 54(B), 55(B), 60(TR), 61(TL), 61(B), 62(TR), 64(TR), 64(B), 68(TR), 69(TL), 80–1, 88(T), 118(T), 119(T), 209(T)

David Nash: 16

Edward Poulton: 12(T), 12(B), 13(B), 14, 15, 17(T), 17(B), 18(T), 18(B), 19(T), 20(T), 21(T), 28(TL), 28(BL), 29(T), 32, 36(T), 40–1, 44(T), 44(B), 45, 50(B), 54(TL), 56(T), 59(B), 60(TL), 61(TR), 62(TL), 64(TL), 68(TL), 69(TR), 72(TR), 75(B), 79, 82(T), 82(B), 83(T), 84(B), 85, 86(T), 88(BL), 92(B), 93(B), 94(B), 95, 96, 98(L), 104(TL), 124(T), 132, 146(TL), 150–1, 154(T), 155(T), 158(TL), 160(TL), 167(B), 170(TL), 192(TL), 197(L), 202(TL), 206(T), 227(B), 228(L), 229(L)

Victor Shreeve: 196(B)

John Tyler: 51(B)

Sidney Woods: 19(B), 21(B), 23, 38–9, 42–3, 208(T)

Figue de Bordeaux (violette)